AF388956

Grid-Connected PV Plants

Grid-Connected PV Plants

Editors

Ángel Molina-García
Rosa Anna Mastromauro

MDPI • Basel • Beijing • Wuhan • Barcelona • Belgrade • Manchester • Tokyo • Cluj • Tianjin

Editors
Ángel Molina-García
Universidad Politécnica de Cartagena
Spain

Rosa Anna Mastromauro
University of Florence
Italy

Editorial Office
MDPI
St. Alban-Anlage 66
4052 Basel, Switzerland

This is a reprint of articles from the Special Issue published online in the open access journal *Energies* (ISSN 1996-1073) (available at: https://www.mdpi.com/journal/energies/special_issues/GC_PVP).

For citation purposes, cite each article independently as indicated on the article page online and as indicated below:

LastName, A.A.; LastName, B.B.; LastName, C.C. Article Title. *Journal Name* **Year**, *Article Number*, Page Range.

ISBN 978-3-03936-848-8 (Hbk)
ISBN 978-3-03936-849-5 (PDF)

Contents

About the Editors

Ángel Molina-García received a degree in electrical engineering from the Universidad Politécnica de Valencia, Spain, in 1998, and a Ph.D. in electrical engineering from the Universidad Politécnica de Cartagena, Spain, in 2003, where he is currently Full Professor. His research interests include wind power generation, PV power plants, energy efficiency, and renewable integration into power systems.

Rosa Anna Mastromauro received M.Sc. and Ph.D. degrees in electrical engineering from the Politecnico di Bari, Bari, Italy, in 2005 and 2009, respectively. Since 2005, she has been with the Power Converters, Electrical Machines, and Drives Research Team, Politecnico di Bari, where she was an Assistant Professor. Currently she is an Associate Professor at the University of Florence, Florence, Italy, and is engaged in teaching courses in power electronics and electrical machines. Her research interests include power converters and control techniques for distributed power generation systems, renewable energies, and transportation applications.

Preface to "Grid-Connected PV Plants"

This Special Issue discusses different aspects of the increasing presence of nonprogrammable renewable energy sources (RESs) in current power systems, mainly focused on photovoltaic (PV) power plants connected to the grid. Under this framework, coordinated voltage and reactive power control analysis are discussed and evaluated, as well as technical and economic PV studies. In addition, some contributions regarding hybrid solutions considering the variable nature of RES and discussion of grid synchronization and PV module orientation are included in this Special Issue, where PV power plant integration is approached from a transversal perspective.

Ángel Molina-García, Rosa Anna Mastromauro
Editors

Article

Optimal Design of Photovoltaic Power Plant Using Hybrid Optimisation: A Case of South Algeria

Tekai Eddine Khalil Zidane [1,*], Mohd Rafi Adzman [1,2], Mohammad Faridun Naim Tajuddin [1], Samila Mat Zali [1], Ali Durusu [3] and Saad Mekhilef [4,5,6]

[1] School of Electrical Systems Engineering, Universiti Malaysia Perlis, Arau 02600, Malaysia; mohdrafi@unimap.edu.my (M.R.A.); faridun@unimap.edu.my (M.F.N.T.); samila@unimap.edu.my (S.M.Z.)

[2] Centre of Excellence for Renewable Energy (CERE), School of Electrical Systems Engineering, Universiti Malaysia Perlis, Arau 02600, Malaysia

[3] Department of Electrical Engineering, Davutpasa Campus, Yildiz Technical University, Istanbul 34220, Turkey; adurusu@yildiz.edu.tr

[4] Power Electronics and Renewable Energy Research Laboratory (PEARL), Department of Electrical Engineering, University of Malaya, Kuala Lumpur 50603, Malaysia; saad@um.edu.my

[5] School of Software and Electrical Engineering, Swinburne University of Technology, Victoria 3122, Australia

[6] Center of Research Excellence in Renewable Energy and Power Systems, King Abdulaziz University, Jeddah 21589, Saudi Arabia

* Correspondence: zidanetekai@hotmail.fr

Received: 5 March 2020; Accepted: 30 April 2020; Published: 1 June 2020

Abstract: Considering the recent drop (up to 86%) in photovoltaic (PV) module prices from 2010 to 2017, many countries have shown interest in investing in PV plants to meet their energy demand. In this study, a detailed design methodology is presented to achieve high benefits with low installation, maintenance and operation costs of PV plants. This procedure includes in detail the semi-hourly average time meteorological data from the location to maximise the accuracy and detailed characteristics of different PV modules and inverters. The minimum levelised cost of energy (LCOE) and maximum annual energy are the objective functions in this proposed procedure, whereas the design variables are the number of series and parallel PV modules, the number of PV module lines per row, tilt angle and orientation, inter-row space, PV module type, and inverter structure. The design problem was solved using a recent hybrid algorithm, namely, the grey wolf optimiser-sine cosine algorithm. The high performance for LCOE-based design optimisation in economic terms with lower installation, maintenance and operation costs than that resulting from the use of maximum annual energy objective function by 12%. Moreover, sensitivity analysis showed that the PV plant performance can be improved by decreasing the PV module annual reduction coefficient.

Keywords: optimal design; photovoltaic power plants; hybrid optimisation; LCOE; PV module reduction

---

1. Introduction

Nowadays, solar photovoltaic energy is being utilised in electrical energy generation to meet the quick-growing consumption and the urgent need for power [1]. Grid-connected photovoltaic (PV) systems with a capacity of 3 kW PV modules could meet the electric demand of a 60–90 m^2 for residential building [2]. By contrast, large-scale PV power plants face some major challenges for the use of vast amounts of components in relation to the cost, reliability, and efficiency, requiring an optimal design of the PV power plant. Recently, the drop in PV module prices of up to 86% from 2010 to 2017 [3] resulted in a decrement in the levelised cost of energy (LCOE) of large-scale PV power plants reaching 0.03 ($) [4].

TRNSYS software has been used to determine the optimum PV inverter sizing ratios [5]. The simulation has been carried out using three types of inverters with low, medium and high efficiency to determine the maximum total output of the PV system. Furthermore, the PV inverter sizing ratio of the grid-connected has been investigated for eight European locations. Mondol et al. suggested that the installation of a PV system with high-efficiency inverter in the sizing of PV and inverter is more flexible than that of a low-efficiency inverter. Artificial intelligence (AI) methods have also been used to optimise the grid-connected PV power plant, as presented in [6], whereas the PV plant global solution is solved through particle swarm optimisation technique (PSO) and compared with a genetic algorithm (GA), based on the total net economic benefit. However, the PSO algorithm showed better performance than the GA approach used in this study. The optimisation design of the grid-connected PV system is introduced in [7]. The decision variables of the proposed methodology are the type of PV modules, inverter, and tilt angle. The study supports the mathematical models of the PV array, inverter and solar irradiance on tilt PV modules surface. The optimisation process considered three types of inverter, four types of PV modules and seven values of tilt angle, as well as the hourly solar irradiance and ambient temperature. As a result, the optimal design of the system is selected based on maximum efficiency.

In 2012 [8], Sulaiman, S.I., et al. proposed a sizing methodology by using an evolutionary programming sizing algorithm. The optimisation procedure supports all possible combinations of PV modules and inverters considering different types of PV modules and inverters. The technical and economic aspects are included in this method, and both the maximum yield factor and the net present value of the PV system were calculated. Chen et al. have proposed an iterative method for the optimal size of inverter for PV systems with maximum savings in nine locations in the USA [9]. The optimisation procedure has selected the gainful inverter size for each location. Additionally, optimum inverter size lower than or the same as that of PV array rated size can be installed, due to the inverter intrinsic parameters, economic and weather considerations. In 2014, Perez-Gallardo proposed an optimal configuration of the grid-connected PV power plant of different PV technologies by using the GA technique, by considering economic, technical and environmental criteria [10]. This study aims to maximise annual energy generation. Another methodology was proposed in [11] to design a PV plant for the self-consumption mechanism for different capacities in the range of 450–1250 kWp for the university campus. The simulation was performed using PV*SOL software.

A study in [12] investigated the selection and configuration of inverter and PV modules for a PV system for minimising costs. The purchasing costs can be reduced by 16.45% of 10 kW by using this model. However, this evaluation model is applicable only at the lowest price and cannot be applied to achieve the highest efficiency in power production. A mathematical procedure is presented in [13–15] to determine the optimal number of rows and a PV module tilt angle for maximising the profit during PV plant lifetime, by considering the effect of shading on the PV module output power. A work in [16] investigated the design of PV systems grid-connected, considering the PV module degradation rate, to select the optimum inverter size for increased energy and reduced cost. Actual inverters with high efficiency offer a wider range than inverter with low-efficiency for sizing factor to increase the energy generation. Research presented in [17] proposed an eco-design for grid-connected PV systems, on the basis of the combination of multi-objective optimisation and other software. The techno-economic and environmental criteria were optimised simultaneously. The installation of thin-film PV modules in PV systems show an advantage over crystalline silicon ones. A methodology for achieving the optimal configuration of large-scale PV power plants to improve its performance is presented in [18]. The optimisation process was performed using different algorithms and is considered to minimise the LCOE by using crystalline silicon and thin-film cadmium telluride PV module technology. According to this study, the proposed technique of grey wolf optimiser showed improved results compared with the other methods in solving the optimal design of the PV power plant. The PV plant LCOE with the thin film had a lower value than crystalline silicon and is more productive. The work reported in [19] proposed a method to convert the design of PV power plants to binary linear programming to achieve an economical design. However, in this method, only the number of inverters and PV modules connected in series and parallel were considered as the design variables. Some other methods have also been employed and published by researchers in this topic to propose a suitable configuration and determine the best solution that considers the environment, economic and technical aspects of the PV system [20–36]. Additionally, references [37–41] reviewed the grid-connected PV system optimisation and challenges.

The average time for the input meteorological data is an essential factor in PV system design, because the monthly and daily average time of the meteorological data fails to determine an optimum design, resulting in the oversizing system and high energy losses and increasing the financial risk of the PV plant. Additionally, the geographic latitude of the PV plant installation site can lead to a significant variation of the PV module optimal tilt angle from one location to another, to convert maximum solar irradiance into electricity and make the PV system more profitable [42].

This paper intends to present a methodology for designing PV power plants by considering semi-hourly time-resolution (i.e., 30 min-average) to address the accuracy of the meteorological data variation, and thus determine the PV plant optimal design and increase its performance. The procedure considers the detailed specifications of the different alternatives of PV modules and inverters to determine the optimum component and system topology for the location under study. Three meteorological parameters of solar irradiation, wind speed, and ambient temperature were measured for 1 year at the installation field are considered. Hybrid grey wolf optimiser-sine cosine algorithm (HGWOSCA) [43] and sine cosine algorithm (SCA) [44] were applied as optimisation techniques to solve the PV plant design problem for two different objectives, including minimum levelised cost of energy (LCOE) and maximum annual energy, while considering many design variables for improving the system performance. The contributions of this article to the book of knowledge in this research field are described below.

- The proposed methodology is suitable to be executed using semi-hourly time-resolution (i.e., 30 min-average) values of the meteorological input data in designing the PV power plant and by introducing an actual PV plant field model, by considering the shape and size of the PV power plant installation area, to arrange all the existing components properly.
- The application of a HGWOSCA optimisation approach after the consideration of two objective functions to design the PV power plant was presented.
- A sensitivity study was performed to investigate the effect of the annual PV module reduction coefficient on PV plant performance.
- A review of the Algerian renewable energy target and its integration was presented.

This paper is organised as follows: Section 2 presents an overview of the renewable energy potential of Algeria, and Section 3 presents the work methodology, including the formulation of the design problem, the PV system description and meteorological data and, the proposed design optimization. In Section 4, the HGWOSCA algorithm is described. Section 5 presents the obtained results with the sensitivity study. Finally, Section 6 presents the conclusions of the paper.

2. The Renewable Energy Potential of Algeria

Algeria has an important potential for electricity generation from renewable energy sources, as performed in several recent studies. However, according to reference [45], approximately 0.415% of electricity in Algeria is generated from renewable energy sources in 2014. The diesel generator is the dominant energy source in rural and Saharan regions in Algeria [46].

2.1. Solar Energy

The potential of solar energy in southern Algeria is the largest in all Mediterranean basins, with 1,787,000 km^2 of Sahara desert, according to the German Aerospace Centre (DLR). The insolation time of almost all the national territory exceeds 2000 h annually and reaches 3900, as shown in Figure 1 (high plains and Sahara) [47]. Over most of the country and during the day, the energy obtained on a horizontal surface of 1 m^2 is nearly 5 kWh or about 2263 kWh/m^2/year in the south and

1700 kWh/m^2/year in the north [48]. This great potential in solar energy compels Algeria to go towards the exploitation of solar energy for power generation, rather than oil and gas. Table 1 shows the rate of sunshine for each region of Algeria.

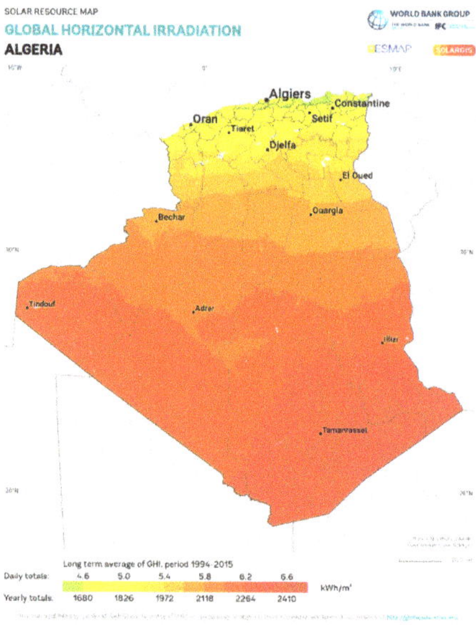

Figure 1. Horizontal Irradiation of Algeria [50].

Table 1. Radiation in Algeria [49].

Region	Coastal	Highlands	Sahara
Area (%)	4	10	86
Average sun hours per year	2650	3000	3500
Energy received kWh/m^2/year	1700	1900	2650

2.2. Other Renewable Energy Potential

Algeria has a considerable average of wind speed that can reach 6 m/s in approximately 50% of the country's surface. The government plans to promote this energy source. Hydroelectricity potential is modest and has little benefit to the Algerian economy. For geothermal energy, the Renewable Energy Development Centre listed more than 200 hot springs, thereby presenting a favourable outlook to exploit this resource [48]. Finally, the biomass potential in Algeria is considerable. It reaches 5 Mtoe/year, but it is still not consumed. This source of energy offers great promises and should be enhanced [51].

2.3. Algerian Renewable Energy Program

Currently, Algeria focuses on the production of electrical energy from renewable sources, especially on solar energy and grid-connected photovoltaic power plants, for a capacity of several megawatts to reduce its dependence on oil rents, thereby representing approximately 96% of export and 17.36% of GDP in 2014 [52]. It also aims to reduce the use of diesel generators, which is the dominant energy source in rural and Saharan regions in Algeria [46].

The development of green energy sources is a key priority for the Algerian government. Nowadays, enhancing the exploitation of renewable energy is a necessity to decrease the amount of CO_2 emissions, considering that Algeria has shown great interest in signing the historic Paris agreement in 2015 [53].

Furthermore, with such a measure, the government can save conventional resources, which are used to generate electricity. The renewable sources showed a poor share in the total energy compared with the conventional sources [54]. The residential electricity sector reached approximately 42% of the total energy consumption [55]. However, the ambitious national renewable energy program allowed one to reach 27% of renewable energy in the national energy mix [56].

Renewable energy sources are the focus of the 2011–2030 development program adopted by the Algerian government to achieve the installation of 22,000 MW of renewables by 2030, including 10,000 MW for export and 12,000 MW for meeting the national market demand [57]. Notably, photovoltaic energy is the dominant renewable source and is expected to reach a capacity of 13,575 MW, representing 62% of the total power installation, as shown in Table 2. In this ambitious program, the government strategy focuses on the development of photovoltaics on a large scale and prepares for the future of Algeria. The adopted program includes the development of wind and CSP energy, biomass, cogeneration and geothermal sources.

Table 2. Phases of the Algerian renewable energy program [58].

Energy Type	1st Phase 2015–2020 (MW)	2nd Phase 2021–2030 (MW)	Total (MW)
Photovoltaic	3000	10,575	13,575
Wind	1010	4000	5010
CSP	-	2000	2000
Cogeneration	150	250	400
Biomass	360	640	1000
Geothermal	5	10	15
Total	4525	17,475	22,000

2.4. Photovoltaic Power Plants Installed in Algeria

In August 2019, the Algerian company for electricity and gas (Sonelgaz) signed an agreement with five companies to construct nine PV power plants in the southern big Sahara region, with a total capacity of 50 MW [59]. This project aims to make a hybrid energy system with an existing gas turbine and diesel generator. The PV plants are installed in different locations, as shown in the following Table 3:

Table 3. Projects of Photovoltaic power plant installation in Algeria.

N	PV Power Plant	Province	Power (MWp)
1	In Guezzem	Tamanrasset	6
2	Tinzaouatine	Tamanrasset	3
3	Djanet	Illizi	4
4	Bordj Omar Dris	Illizi	3
5	Bordj Badji Mokhtar	Adrar	10
6	Timiaouine	Adrar	2
7	Talmine	Adrar	8
8	Tabelbala	Bechar	3
9	Tindouf	Tindouf	11

Several large-scale PV power plants have been installed and connected to the electric grid in different locations across the country, and most of the PV plants are located in the big Sahara in the South, thereby indicating that Algeria benefits from this source of energy, as shown in Table 4. In addition to the absence of batteries that reduce the total capital cost, this system allows power generation surplus in terms of consumption of the load to be automatically injected into the electric grid.

Table 4. Photovoltaic power plants installed in Algeria.

N	PV Power Plant	Province	Power (MWp)	Energy Production (GWh)-June 2017	Commissining Date	Area (km^2)	Distance to Electric Grid (km)	Voltage Level at Point of Coupling (kV)	PV Modules Type	Topology
1	Adrar	Adrar	20	59.585	Oct-15	0.4	2.8	30	Poly	C
2	Kabertène	Adrar	3	9.584	Oct-15	0.06	0.2	30	Poly	C
3	In Salah	Tamanrasset	5	12.328	Feb-16	0.1	0.5	30	Poly	C
4	Timimoune	Adrar	9	23.822	Feb-16	0.18	9	30	Poly	C
5	Regguen	Adrar	5	12.221	Jan-16	0.1	0.22	30	Poly	C
6	Zaouiat Kounta	Adrar	6	15.213	Jan-16	0.12	0.24	30	Poly	C
7	Aoulef	Adrar	5	12.557	Mar-16	0.1	0.4	30	Poly	C
8	Tamanrasset	Tamanrasset	13	36.41	Nov-15	0.26	8.8	30	Poly	C
9	Djanet	Ilizi	3	10.729	Feb-15	0.06	0.35	30	Poly	C
10	Tindouf	Tindouf	9	6.376	Dec-15	0.18	0.3	30	Poly	C
11	Oued Nechou	Ghardaia	1.1	4.593	Jul-14	0.1	0.4	30	m-Si,c-Si,Thin film, si-amorphe	C
12	Sedret Leghzel	Naâma	20	40.751	May-16	0.42	1.3	60	Poly	C
13	Oued El Kebrit	Souk Ahras	15	28.9	Apr-16	0.32	6.5	30	Poly	C
14	Ain Skhouna	Saida	30	14.213	May-16	0.42	12	60	Poly	C
15	Ain El Bel 1	Djelfa	20	25.134	Apr-16	0.4	3.9	60	Poly	C
16	Ain El Bel 2	Djelfa	33	25.134	Apr-16	0.8	3.9	60	Poly	C
17	Lekhneg 1	Laghouat	20	53.576	Apr-16	0.4	12	60	Poly	C
18	Lekhneg 2	Laghouat	40	53.576	Apr-16	0.8	12	60	Poly	C
19	Telagh	Sidi-Bel-Abbes	12	7.417	Sep-16	0.3	6	60	Poly	C
20	Labiodh Sidi Chikh	El-Bayadh	23	19.146	Nov-18	0.392	0.9	60	Poly	C
21	El Hdjira	Ouargla	30	9.738	Jul-16	0.6	1	60	Poly	C
22	Ain-El-Melh	M'Sila	20	16.473	Sep-17	0.4	0.9	60	Poly	C
23	Oued El Ma	Batna	2	-	Mar-17	0.3	0.5	30	Poly	C

3. Methodology

3.1. Formulation of the Design Problem

The single objective optimisation function is used to find the optimum solution corresponding to the minimum or maximum value defined by the objective function. In contrast, multi-objective optimisation combines two or more individual objective functions to determine a set of trade-off solutions, which allow decision makers to select the most suitable solution based on the problem requirements [60]. In this study, in sizing optimisation methodology depending on the requirements of the power plant designer, each of the two objectives can be used to produce an optimal design for the PV power plant. In addition, for comparison purposes, the optimum values are calculated by using each objective function individually to evaluate the PV power plant performance.

Furthermore, multi-objective optimisation can be used in the design of PV systems with a small capacity in the range of kW, with a small number of PV modules and inverters or in hybrid renewable energy systems for example (PV-wind) or (PV-diesel denerator-battery). However, in large scale PV power plants (i.e., >200 kW nominal power rating—the largest plants reaching several tens of MW of capacity), with a considerable number of components required in PV plant installation, it is well-known that the levelised cost of energy (LCOE) is applied to enable the reduction of the PV plant cost per watt of nominal power that is installed [61,62], for this reason, single objective optimisation is used. Additionally, a recent study is presented in [63] to investigate the LCOE of large scale PV power plants at 8 PV plants ranging from 1 to 46 MWp and many similar studies can be found in the literature.

In this section, two objective functions are considered to evaluate the PV power plant performance and to solve its complex design problem. The design variables and constraints of the proposed methodology are also explained.

3.1.1. Objective Function

In this work, the LCOE and maximum annual energy were set as objective functions to determine the optimal solution of the PV plant design. These two objective functions can be combined to form a single optimisation function.

The first part presents the LCOE which is calculated on the basis of the sum of maintenance, operation and installation costs of the plant divided by the total energy generation of the plant during its lifetime. The LCOE method is generally applied to compare power plants with different energy generation sources, by considering the appropriate cost structures. However, the best LCOE for power plants presents the lowest possible investment with high annual energy production. The second part presents the maximum amount of annual energy that can be captured by the PV modules during the PV plant in its lifetime, which is 25 years. The single optimisation function is expressed by the following equation:

$$\frac{min}{X}\left[\left(\frac{C_c(X) + C_M(X)}{E_{tot}(X)}\right)\cdot a - \left((1-a)\cdot\left(P_{plant}(X)\cdot n_s\cdot EAF\right)\right)\right] \quad (1)$$

where n_s is equal to 1 year.

The optimum values are calculated by using each objective function individually. In other words, in the objective function, a is a binary number; if a is equal to 0, the target of the objective function is maximum energy and, if a is equal to 1, the objective function target is minimum LCOE.

3.1.2. Design Variables

The proposed optimisation algorithm was used for the calculation of all the decision variables, to determine the optimum design of the PV power plant. The chosen optimisation algorithm should have high performance in determining the best design variables and solving the design problem. In this methodology, the proposed decision variables, including the number of PV modules connected in series (N_s) and parallel (N_p), number of PV module lines per row (N_r), the distance between two adjacent rows (F_y), the tilt angle of the PV module (β), the orientation of PV modules (PV_{orien}), that can be installed vertically or horizontally, optimum PV module (PV_i), and inverter (IN_i), can be selected on the basis of several alternatives from a list of possible candidates.

The vector of the decision variables are summarized as given by the following expression:

$$X = \begin{bmatrix} N_s & N_p & N_r & \beta & F_y PV_{orien} & PV_i & IN_i \end{bmatrix} \tag{2}$$

3.1.3. Constraints

During the design of the PV power plant, many constraints are considered to account for the limits of the different parameters of the whole system. The following expression shows the limitation of some variables:

$$N_{s,min} \leq N_s \leq N_{s,max} \tag{3}$$

$$1 \leq N_p \leq N_{p,max} \tag{4}$$

$$1 \leq N_r \leq N_{r,max} \tag{5}$$

$$0 \leq \beta \leq 90 \tag{6}$$

$$S_{occupied} \leq Avalaiblearea \tag{7}$$

The following equality constraint expressions were used to select the PV module and inverter from the list of candidates:

$$PV_1 + PV_2 + \ldots = 1 \tag{8}$$

$$INV_1 + INV_2 + \ldots = 1 \tag{9}$$

3.2. System Description and Meteorological Data

This paper focuses on the optimal design of large-scale PV power plants connected to the electric grid. The optimisation process was carried out using HGWOSCA and SCA techniques. The proposed methodology supports PV plant configurations for both central and string topologies and the arrangement of the components within the installation area. Moreover, it considers a list with different types of PV modules and inverters and their specifications as candidates to design the PV power plant, as illustrated in Tables 5 and 6. The actual meteorological data, such as solar irradiance, wind speed, and ambient temperature, are also considered. The optimisation process in this methodology aims to determine the number of PV modules connected in series and parallel, the number of PV module lines per row, the distance between two adjacent rows, the tilt angle of the PV module, the orientation of PV modules that can be installed vertically or horizontally, the optimum PV module and inverter, from a list of possible candidates. Two objective functions, namely the minimum LCOE and maximum annual energy, were considered. Furthermore, the effect of the annual PV module reduction coefficient on PV plant performance was determined.

Table 5. Photovoltaic (PV) modules specifications at standard test condition.

Specification at STC	Unit	PV1	PV2	PV3
Nominal maximum power ($P_{mpp,stc}$)	W	280	285	295
Optimum operating current ($I_{mpp,stc}$)	A	8.84	9.02	9.08
Optimum operating voltage ($V_{mpp,stc}$)	V	31.7	31.6	32.5
Current temperature coefficient (K_i)	(%/C)	0.05	0.05	0.05
Voltage temperature coefficient (K_v)	(%/C)	−0.29	−0.32	−0.29
Open circuit voltage ($V_{oc,stc}$)	V	38.4	38.3	39.6
Wind speed temperature coefficient (K_r)	-	1.509	1.4684	1.509
Length ($L_{pv,1}$)	m	1.65	1.65	1.65
Width ($L_{pv,2}$)	m	0.992	0.992	0.992
Efficiency	%	17.1	17.4	18
Type	-	Mono	Poly	Mono

Table 6. Inverters specifications at standard test condition.

Specification	Unit	INV1	INV2	INV3
Nominal power (P_i)	kW	50	7	500
Minimum input voltage ($V_{i,min}$)	V	250	335	450
Maximum input voltage ($V_{i,max}$)	V	950	560	880
Maximum power point tracking voltage ($V_{i,mppt,max}$)	V	850	540	820
Power loss ($P_{i,sc}$)	W	1.5	1	200
Efficiency (η_{inv})	%	97.5	95.3	98.7

3.2.1. System Description

The architecture of PV plants is composed of several hundreds of PV modules that produce DC power depending on the meteorological parameters (solar radiation and temperature) [64] and inverters that permit the conversion from DC to AC power and ensure the maximum power extraction from the PV modules [65]. The generated power is injected directly into the electric network at the point of common coupling by using step-up transformers [66].

In PV plants, PV modules are connected in series (N_s) to form a string. These strings are rationalised and connected in parallel ($N_p \geq 1$) to form a PV array. In the case of central inverters topology, several hundreds of PV modules are connected to one inverter, and junction boxes need to be used through the DC main cable before leaving for the inverter. In string inverter topology, one string is connected directly to one inverter, and junction boxes need not be placed in the installation field.

The PV arrays within the available area are arranged in multiple rows. Each row consists of multiple lines of PV modules. The number of lines per row is equal to N_r. Considering the shading effect, the inclined adjacent rows are installed with space in between. The tilt angle (β) of PV modules is constant during the PV plant lifetime.

The universal transverse Mercator coordinate (UTM, X: east, Y: north) system was used to model the PV power plant area. Several PV modules vary from one row to another, as the row length varies on the basis of the installation area shape.

3.2.2. Selected Site

The Djanet village is located in the South East of Algerian Sahara in the province of Illizi, and it is characterised by a hot desert climate according to longitude 9.28° E, latitude 24.15° N and an altitude of 1030 m. Djanet village electrification is still dominated by diesel generation (DG). The delivery of fuel leads to an increase in the cost and the maintenance of the units. However, a PV plant grid connected with a capacity of 3 MW was installed in 2015, and its extension is expected to reach 7 MW to supply the village and decrease the units of DG.

3.2.3. Meteorological Data

The performances of PV modules depend on solar irradiation, ambient temperature, and wind speed. These data have been recorded to step times of semi-hourly and hourly. Solar irradiation is expressed in Watts per square meter (W/m^2). Ambient temperature is in degrees Celsius (°C). Wind speed is in m/s. Figures 2 and 3 show the daily assessment of the meteorological characteristics. The semi-hourly data of solar irradiance and ambient temperature are observed as more accurate than the hourly data. That is because the peak values of the solar irradiance and ambient temperature may not be recorded in the hourly data. This happens because the data only records point values for every hour, while within a period of time the meteorological data may have significant fluctuation. Thus, semi-hourly data is more precise and accurate compared to hourly data measurements.

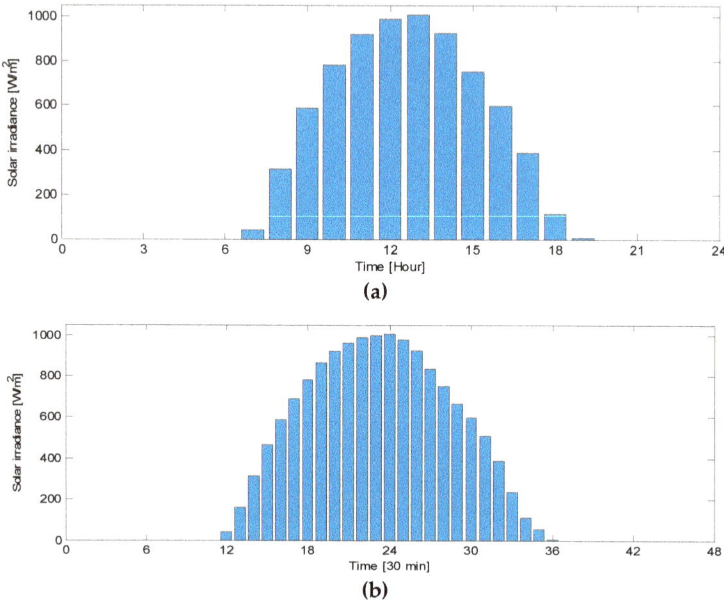

Figure 2. Irradiance data. (**a**) Hourly average time. (**b**) 30 min average time.

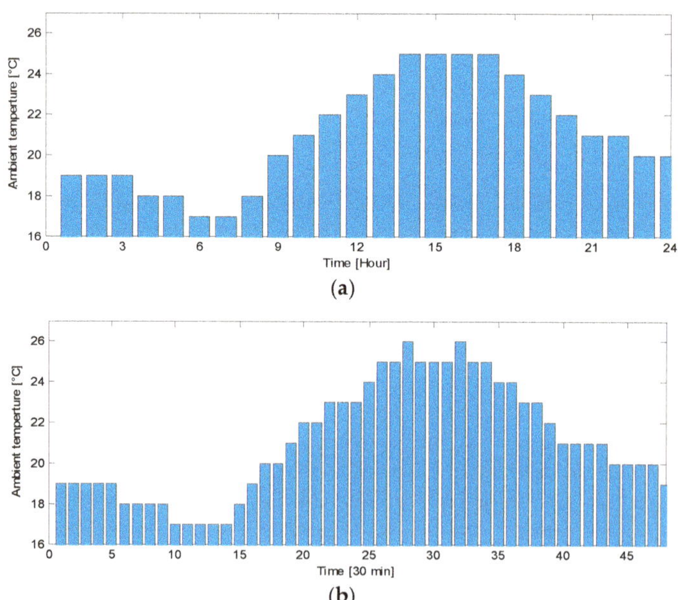

Figure 3. Temperature data. (**a**) Hourly average time. (**b**) 30 min average time.

To determine the energy generated by the PV power plant, knowledge of the solar irradiance profile during the year is required. The solar irradiation for hourly and semi-hourly (i.e., 30 min-average) data of the selected location are plotted in Figures 4 and 5. As we can see, the radiation intensity remained high over the year. It is observed that the maximum value of solar irradiance is reached in

March. As for the minimum value of irradiance, this is recorded in June. Each vector represents the same size for 1 year, where the one for semi-hourly step is equal to 17,520 data and the one for hourly step is equal to 8760 data. The climatic conditions of the location are as follows: high solar irradiance potential; ambient temperature with a maximum average of (29.7 °C) in summer; and the sky is mostly clear during the whole year.

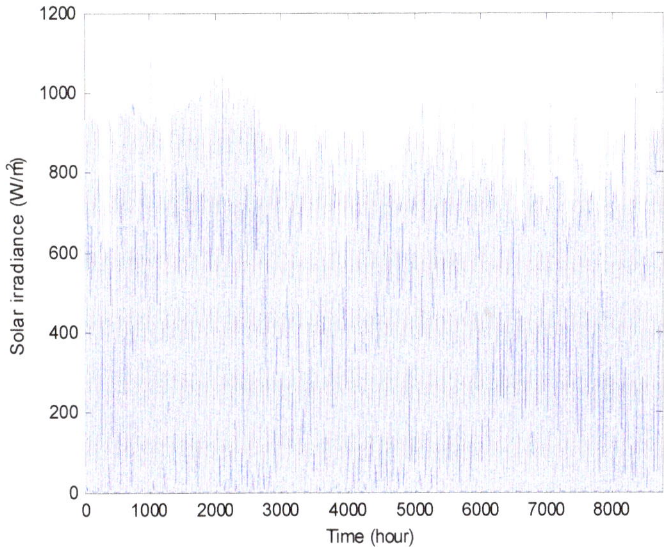

Figure 4. Hourly solar irradiance over the year (W/m^2).

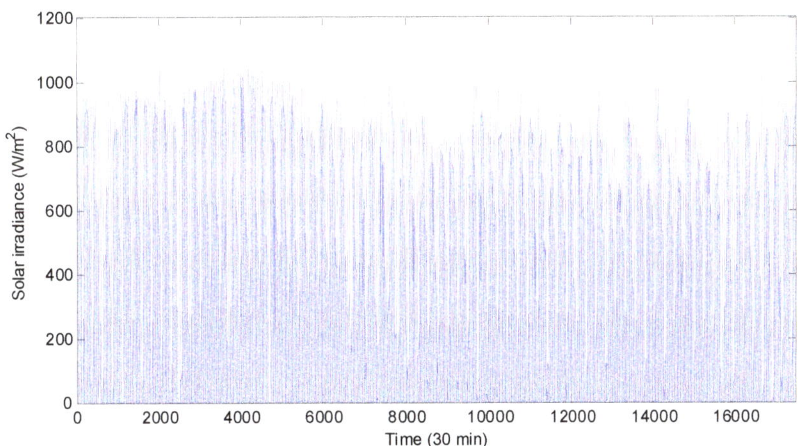

Figure 5. Solar irradiance over the year (W/m^2).

3.3. Proposed Design Methodology

The proposed methodology was applied to solve the PV power plant design and aimed to determine the optimal sizing and configuration of the PV plant, as shown in Figure 6. In the following section, the PV plant design parameters were calculated step by step by considering the measured meteorological data of the location, PV modules, inverter specifications, area coordinates, and cost units.

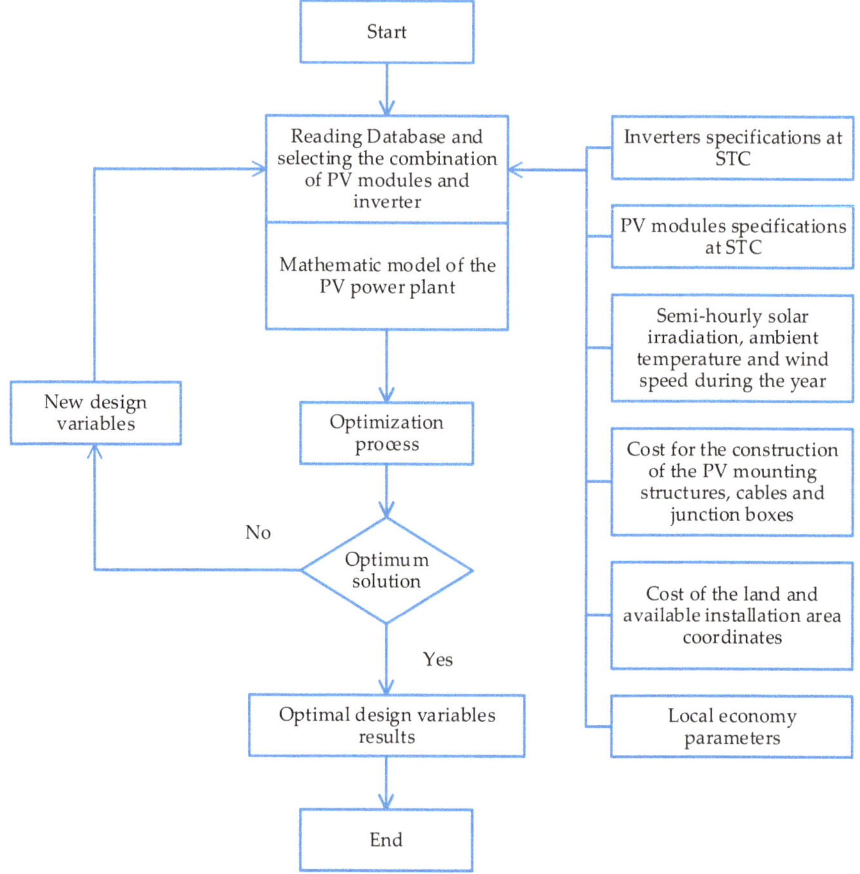

Figure 6. Flowchart of the proposed design methodology.

3.3.1. Irradiance Model

Solar irradiance on tilted PV modules surface is a very important factor in the optimal design of PV power plants. Installation areas with PV modules facing south are suitable for PV power plants [35]. Additionally, a good agreement was shown by PV plants oriented towards the south by using the isotropic model [67].

Several models, which are classified as isotropic and anisotropic, can be used to estimate the solar irradiance on a tilted plane [67,68]. The treatment of diffuse radiation was the only difference between these two models, while the rest is treated the same. However, the isotropic model developed by Liu and Jordan [69] was applied to this methodology to estimate the solar irradiance on the tilted PV module surface [67].

The total radiation received on a horizontal surface (global radiation: I) can be divided into two components: beam and diffused radiation. The estimation of solar radiation on the tilted surface is calculated on the basis of these two components. The total radiation received on the horizontal surface is given by the following equation:

$$I(t) = I_b(t) + I_d(t) \tag{10}$$

The index of transparency of the atmosphere or the clearness index k_T of the sky is an essential factor. The clearness index is the function of the ratio between the extraterrestrial and horizontal radiation, as expressed by the following equation:

$$k_T(t) = \frac{I(t)}{I_0(t)} \tag{11}$$

The diffuse fraction of total horizontal radiation depends on the clearness index of the sky [67] and is expressed by the following equation:

$$\frac{I_d(t)}{I(t)} = \begin{cases} 1.0 - 0.09k_T(t), k_T(t) \leq 0.22 \\ 0.9511 - 0.1604k_T(t) + 4.388(t)k_T^2 - 16.638k_T^3(t) + 12.336k_T^4(t), \ 0.22 < k_T(t) \leq 0.8 \\ 0.165, k_T(t) \leq 0.80 \end{cases} \tag{12}$$

manipulating Equation (10), the beam radiation is given by the following expression:

$$I_b(t) = I(t) - I_d(t) \tag{13}$$

The total incident solar radiation on tilted surface is the sum of three components, namely, beam radiation from direct radiation of the inclined surface, diffuse radiation and reflected radiation.

$$I_T(t, \beta) = I_B(t) + I_D(t) + I_R(t) \tag{14}$$

The beam irradiance on an inclined surface can be calculated on the basis of multiplication between beam horizontal irradiance and beam ratio factor R_b, as shown in the following expression:

$$I_B(t) = I_b(t)R_b(t, \beta) \tag{15}$$

where the beam ratio factor R_b is a function of the ratio between beam irradiance on the inclined surface and horizontal irradiance, as expressed in the Equation (18).

The first component is the incidence angle $\cos(t, \beta)$, which can be derived as follows:

$$\begin{aligned} \cos(t, \beta) \quad &= \sin \delta(t) \sin \varphi \cos \beta - \sin \delta(t) \cos \varphi \sin \beta \cos \gamma \\ &+ \cos \delta(t) \cos \varphi \cos \beta \cos \omega(t) \\ &+ \cos \delta(t) \sin \varphi \sin \beta \cos \gamma \cos \omega(t) \\ &+ \cos \delta(t) \sin \beta \sin \gamma \sin \omega(t) \end{aligned} \tag{16}$$

where δ is the solar declination angle, φ is the location latitude, γ is the surface azimuth angle, and ω is the hour angle. The global radiation on the inclined surface calculation model's error was lower than 3% [70].

The second component deals with solar zenith angle $\cos \theta_z$ and can be calculated using the following equation:

$$\cos \theta_z(t, \beta) = \cos \gamma \cos \delta(t) \cos \omega(t) + \sin \varphi \sin \delta(t) \tag{17}$$

$$R_b(t, \beta) = \frac{\cos(t, \beta)}{\cos \theta_z(t, \beta)} \tag{18}$$

Diffuse irradiance on an inclined surface is computed on the basis of the isotropic sky model. A well-known isotropic model was introduced by Liu and Jordan (1963). This model is simple, and the diffuse radiation has a uniform distribution over the skydome. The diffuse radiation on the inclined surface increases with an increasing amount of seen by the inclined surface, as expressed in Equation (19).

$$I_D(t) = I_d(t)\left(\frac{1 + \cos\beta}{2}\right) \tag{19}$$

where β is the surface tilt angle and considered as a design variable. Its optimal values are computed by the optimisation algorithm.

The reflected irradiance on an inclined surface is expressed by Equation (20) and depends on the transposition factor for ground reflection R_r given by Equation (21) and the reflectivity of the ground ρ that is equal to 0.2 [68].

$$I_R(t) = I(t)\rho R_r \tag{20}$$

$$R_r(t) = \frac{1 - \cos\beta}{2} \tag{21}$$

3.3.2. Area Calculation Model

In actual cases, the PV power plant installation area is limited in surface and does not have a uniform shape. However, this proposed methodology can be applied to all actual area shapes to determine the optimal size and configuration of large-scale PV power plants. This methodology supports actual area shapes by using the coordinates of the location under study. Additionally, the universal transverse Mercator coordinate (UTM, X: east, Y: north) system is used to model the PV power plant area. Furthermore, the PV plant occupying the surface, length, and width of each row, junction boxes, and cable length are computed on the basis of the coordinates. As mentioned in the previous section, PV nodules are oriented towards the south in the installation field, as illustrated in Figure 7.

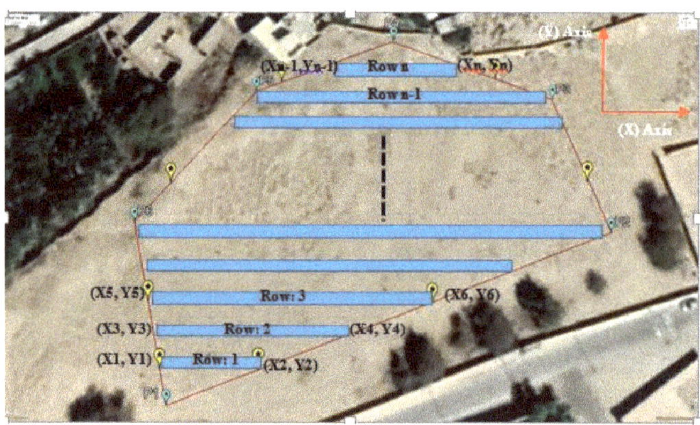

Figure 7. Arrangement of PV modules according to PV plant shape.

The Y-axis was used to calculate the total number of rows supported by the PV plant area and the width (W_T). of each row. The space between two adjacent rows (F_y), which is a design variable, and its optimum value were calculated by the algorithm process. The following equation expresses the total number of rows:

$$N_{row} = floor\left(\frac{\max(Y) - \min(Y)}{W_T + F_y}\right) \tag{22}$$

After calculating the total number of rows (N_{row}), it can be used to calculate the north coordinates of each row (Y_{Ni}) in the PV plant, as expressed in the following equation:

$$Y_{Ni} = \min(Y) + \left(W_T + F_y\right)N_i \tag{23}$$

The parameters (N_i) and (Y_{Ni}) are subject to a constraint in this methodology, as provided in Equations (24) and (25):

$$1 \le N_i \le N_{row} \tag{24}$$

$$min(Y) < Y_{Ni} < max(Y) \tag{25}$$

The X-axis presents the east coordinates and is used for calculating the length of each row in the PV plant X_{Ni}, as expressed in the following straight-line equation:

$$X_{Ni} = \frac{(X_2 - X_1)(Y_{Ni} - Y_1)}{Y_2 - Y_1} + X_2 \tag{26}$$

where (X_1, Y_1). and (X_2, Y_2) correspond to the coordinates of two consecutive points. The parameter X_{Ni} is a constraint in this methodology, as provided in the following expression:

$$min(X) \leq X_{Ni} \leq max(X) \tag{27}$$

The row length (M_{row_i}) is obtained after the calculation of the east coordinates (X_{Ni}) of each row, considering the difference between these coordinates, and is expressed by using Equation (28):

$$M_{row_i} = X_{Ni1} + X_{Ni2} \tag{28}$$

The PV power plant area calculation process considers other important parameters, such as row height (H_T), row width (W_T) and the space between two adjacent rows (F_y). These parameters can be calculated on the basis of the following equations:

$$W_T = N_r L_{pv,2} \cos \beta \tag{29}$$

$$H_T = N_r L_{pv,2} \sin \beta \tag{30}$$

$$F_y = dH_T \tag{31}$$

where (N_r) and (F_y) are considered as design variables, and their optimal values are calculated via optimisation. Notably, (N_r) is the number of PV module lines in each row, and (F_y) is the distance between two adjacent rows. In this methodology, all rows in the installation area have the same lines of PV modules. The arrangement of rows and PV modules in a row within the installation area is shown in Figure 7.

3.3.3. Components Arrangement

The arrangement of the components within the installation area is an essential part of the PV plant design process in the presence of several parameters, such as the location characteristics and the device's specifications. In addition, component arrangement depends on the optimal topology selected by the optimisation algorithm. Furthermore, the distribution of a large amount of the components among the PV power plant is computed in terms of several constraints.

However, PV modules and inverters are the two main devices considered in the PV power plant arrangements. Additionally, in case of the optimisation algorithm select central topology, the junction box arrangement is considered, and its distribution among the PV modules and the inverters is calculated on the basis of its rating power.

Finally, the PV power plant device arrangement is influenced by the amount of solar irradiance, ambient temperature, wind speed, and the geographic location. These parameters affect the tilt angle of PV modules and increase or decrease the PV module energy output, leading to the installation of varying numbers of inverters in the PV plant. Moreover, in this methodology, the aforementioned parameters are considered to control the total cost.

Dependent on PV inverter size, the number of series PV modules in each string (N_s) and parallel PV modules (N_p) should be computed by the algorithm to meet a specific voltage and current requirement of inverters. On the one hand, to avoid the inverter damage that can be caused by overvoltage in case of low temperature in some locations, in every string, the number of PV modules connected in series has to be optimally computed. On the other hand, the number of parallel-connected PV modules (N_p) multiplied by its current is equal to the input current of the inverter. To avoid the inverter damage created by the overcurrent locations with high solar irradiance, a limited number of PV modules connected in parallel (N_p) should be addressed.

The first part handles PV modules distribution among the inverters and their arrangement within the PV plant area. The number of series (N_s) and parallel (N_p) PV modules are computed in accordance with the optimum selected inverter by the optimisation process. In this proposed methodology, the number of PV modules connected in series (N_s) and parallel (N_p) were considered as the design variables, and their optimum values were calculated using the optimisation algorithm. The (N_s) design variable involves a number of minimum $(N_{s,min})$ and maximum $(N_{s,max})$ PV modules, and these limitations can be calculated on the basis of the inverter input voltage range in [11,22], as expressed in the following equations:

$$N_{s,min} = \frac{V_{i,min}}{V_{mpp,min}}$$

(32)

$$N_{s,min} = \frac{V_{i,max}}{V_{oc,max}}$$

(33)

$$N_{sm,2} = \frac{V_{i,mpptmax}}{V_{mpp,max}}$$

(34)

$$N_{s,max} = \begin{cases} N_{sm,1}, & N_{sm,1} \leq N_{sm,2} \\ N_{sm,2}, & N_{sm,2} < N_{sm,1} \end{cases}$$

(35)

The maximum number of PV modules connected in parallel (N_p) was calculated according to the selected inverter by using the nominal power (P_i), and the PV module maximum output power $(P_{mpp,max})$ was selected with respect to the optimum number of PV modules connected in series (N_s) [22], as provided in the following expression:

$$N_{p,max} = \frac{P_i}{N_s P_{mpp,max}}$$

(36)

As mentioned in the previous section, the arrangement of PV modules in the PV plant area requires the use of the length of each row in the PV plant to determine the optimum number of PV modules installed in each line (N_{c_i}) and the total number in each row $(N_{row_i,\,pv})$. The total number of PV modules installed in each line (N_{c_i}) of rows, which are described as the function ratio between the length of each row $\left(M_{row_i}\right)$ and the length of the optimum PV modules $\left(L_{pv,1}\right)$, is given in the following equation:

$$N_{c_i} = \frac{M_{row_i}}{L_{pv,1}}$$

(37)

The total number of PV modules installed in each row $\left(N_{row,pv}\right)$ depends on the number of PV module lines (N_r), which is a design variable in this methodology, and its optimum value is computed by the optimisation algorithm.

$$N_{row_{i,pv}} = N_r N_{c_i}$$

(38)

The sum of PV modules in each row of the PV plant results in their total number in the installation area as expressed in the following equation:

$$N_{\mathrm{I}} = \sum_{1}^{i} N_{row_{i,pv}}$$

(39)

The number of series (N_p) and parallel (N_p) PV modules are the main parameters in the inverter calculation process. These design variables determine the number of blocks, and x_{inv} represents the pieces of inverters in blocks, and each piece is composed of N_{block} [22], as given in the following equations:

$$N_{block} = N_s N_p \tag{40}$$

$$y = (N_i\,,N_{block}) \tag{41}$$

$$x_{inv} = \frac{N_i - y}{N_{block}} \tag{42}$$

Finally, the total number of inverters is calculated on the basis of the following expression:

$$N_i = \begin{cases} x_{inv},\left(\frac{y}{x_{inv}}\right)P_{pv,stc} \leq 0.1P_i \\ x_{inv} + 1,\left(\frac{y}{x_{inv}}\right)P_{pv,stc} > 0.1P_i \end{cases} \tag{43}$$

3.3.4. PV Plant Total Energy

The proposed methodology offers many alternatives for PV modules with different specifications. Additionally, the optimisation algorithm was applied to determine the best candidate for the design of the PV plant and the optimum configuration of the PV plant as a global solution. However, the PV module output power depends on the amount of solar radiation, ambient temperature, wind speed, and electrical characteristics. Moreover, a recent review [71] has covered approximately 70 important papers on PV cell modelling, and the equations used in this proposed methodology have been applied in several papers, as shown in this review. The equations have been used in a recent paper [72], and the obtained results by the proposed procedure are more accurate than the [73] model, which involves the use of the same equations. Accordingly, these equations are suitable for calculating the performance of PV modules in our proposed design procedure.

The PV power plant consists of a large number of PV modules. Additionally, the output power is assumed to be the same for all PV modules in the PV plant, except for the southernmost row, which is considered never shaded. More importantly, the degradation of PV modules is inevitable regardless of the size of a PV power plant [74,75]. However, this research considered the PV module output power derating factor (d_f) due to soiling effect on the PV module surface, which is equal to $d_f = 0.069$, and the annual reduction coefficient r of PV module [34], which is equal to 0.5%. Finally, PV modules output power can be calculated using the following expression:

$$P_{pv}(t,\beta) = (1 - r)\left(1 - d_f\right)P_{mpp}(t,\beta) \tag{44}$$

where P_{mpp} presents the produced power by each PV module in the PV plant.

The produced energy can be affected by the shadow area on PV modules and is related to the shade impact factor (SIF) [76], and its value is equal to 2 [35]. This parameter can be obtained using the following equation:

$$A_{S_i}(t) = \xi_i(t)SIF \tag{45}$$

where $(\xi_i(t))$ presents the ratio of the shadow area.

The total energy of the PV power plant can be calculated according to the optimum inverter topology selected by the optimisation algorithm. Furthermore, the PV power plant produced energy and the total cost can be influenced by the selected inverter topology. For string inverter topology, the following equation is applied to calculate the PV plant output power:

$$P_{plant}(t,\beta) = n_{tr}(1 - \eta_{cac})(1 - \eta_{cic})P_{o_i}(t,\beta)N_i \tag{46}$$

where (P_{o_i}) is the inverter output power, (N_i) represents the total number of inverters, (n_{tr}) is the transformer efficiency, (η_{cac}) presents the AC cable losses and (η_{cic}) is the interconnection cable losses.

In the case of central inverter topology, PV plant output power can be obtained using the following equation:

$$P_{\text{plant}}(t, \beta) = (1 - \eta_{cdc})n_{mppt}n_{inv}n_{tr}(1 - \eta_{cac})(1 - \eta_{cic})\sum_{1}^{row_i} P_{row_i}(t, \beta) \tag{47}$$

where $P_{row_i}(t, \beta)$ presents the PV row output power, n_{mppt}, n_{inv} and n_{tr} are the efficiencies of the PV module, inverter and transformer, respectively, and η_{cdc} and η_{cac}, are the DC and AC cable losses, respectively.

However, in this methodology, the PV plant energy generation was directly injected to the electric network over its operational lifetime, and it was calculated using Equation (48):

$$E_{tot} = P_{\text{plant}}(t, \beta)n_s EAF \tag{48}$$

where EAF is the energy availability factor, and (n_s) is the PV plant operational lifetime.

3.3.5. PV Plant Total Cost

The PV power plant consists of two types of costs, as expressed by Equation (49):

$$C_{tot} = C_c + C_M \tag{49}$$

The installation cost (C_c) deals with the cost of the device, such as C_{pv}, C_{inv} which represents the unit cost of the PV modules and inverters, respectively. In addition, C_B is the PV module mounting structure cost. Moreover, C_{cb}, C_{tr}, C_{pd}, and C_{cm} represent the costs of the cable, transformer, protection devices and monitoring system, respectively. Finally, C_L represents the cost of the plant area. The installation cost is expressed in Equation (50):

$$C_c = N_I C_{pv} + N_i C_{inv} + C_L + C_B + C_{cb} + C_{tr} + C_{pd} + C_{cm} \tag{50}$$

The operation and maintenance costs of the PV plant during its lifetime depend on the annual inflation rate (g), the nominal annual interest rate (i_r). and the operation and maintenance costs per watt (M_{op}), as given in the following expression:

$$C_M = N_I P_{pv,stc} M_{op}(1 + g)\left[\frac{1 - \left(\frac{1+g}{1-i_r}\right)^{n_s}}{i_r + g}\right] \tag{51}$$

4. Hybrid Grey Wolf Optimizer-Sine Cosine Algorithm (HGWOSCA)

SCA and grey wolf optimiser (GWO) are meta-heuristic optimisation algorithms recently developed by Mirjalili et al. [44,77]. Both SCA and GWO approaches show high performance compared with other well-known meta-heuristic algorithms [44,77]. The hybrid GWO-SCA technique was introduced by N. Singh et al. [43] for combining the advantages of both approaches. In the GWO-SCA hybrid approach, GWO presents the main part, whereas the implementation of SCA assists in the optimisation of GWO. An improvement in the position, speed, and convergence of the best grey wolf individual alpha (α) by using the original equation expressed in [77], is achieved by applying the position updating equations of the SCA approach, as illustrated in [44].

The position of the current space agent is updated on the basis of the following equation:

$$\vec{x}_2 = \vec{x}_\beta - \vec{a}_2 \cdot (\vec{d}_\beta), \vec{x}_3 = \vec{x}_\delta - \vec{a}_3 \cdot (\vec{d}_\delta) \tag{52}$$

where $\vec{a}$ is random value in the gap [−2a, 2a].

The position of $\vec{X}_\beta$, $\vec{X}_\delta$. and $\vec{X}_\alpha$. is updated using the following equation:

$$\frac{\vec{x}_1 + \vec{x}_2 + \vec{x}_3}{3} \tag{53}$$

Details and description of the HGWOSCA approach can be found in reference [43]. Furthermore, the computational procedure of the HGWOSCA approach is illustrated in Figure 8.

1.	**Begin of Algorithm**
2.	Initialize the grey wolf population xi = 1,2,.....,n
3.	Initialize the parameters A, a and C
4.	Calculate the fitness of each search agent
5.	$\overrightarrow{X_\alpha}$= the best search agent
6.	= the second best search agent
7.	$\overrightarrow{X_\delta}$ = the third best search agent
8.	While t < Max_generation
9.	For search space
10.	Update the position of the current space agent in the basis of equation (53)
11.	End
12.	Update the parameters a, A and C
13.	Calculate the fitness of search agent
14.	Update the $\overrightarrow{X_\beta}$, $\overrightarrow{X_\delta}$ by equation (52) and $\overrightarrow{X_\alpha}$ as below
15.	if rand() < 5
16.	then
17.	$\overrightarrow{D_\alpha} = \mathbf{rand}() \times \mathbf{sin}(\mathbf{rand}()) \times + \mid \overrightarrow{C_1}.\overrightarrow{X_\alpha} - \vec{X} \mid$
18.	else
19.	$\overrightarrow{D_\alpha} = \mathbf{rand}() \times \mathbf{cos}(\mathbf{rand}()) \times + \mid \overrightarrow{C_1}.\overrightarrow{X_\alpha} - \vec{X} \mid$
20.	$\overrightarrow{X_1} = \overrightarrow{X_\alpha} - \overrightarrow{A_1}.\overrightarrow{D_\alpha}$
21.	end if
22.	end else
23.	end while
24.	Return $\overrightarrow{X_\alpha}$

Figure 8. Pseudo-code of the hybrid grey wolf optimiser-sine cosine algorithm (HGWOSCA).

Although the GWO and SCA are able to expose an efficient accuracy in comparison with other well-known swarm intelligence optimisation techniques, it is not fitting for highly complex functions and may still face the difficulty of getting trapped in local optima [43]. Thus, a new hybrid variant based on GWO and SCA is used to solve recent real-life problems.

5. Results and Discussion

The proposed methodology has been implemented in MATLAB software and applied to the development of the optimal design of a PV plant connected to the electric grid. Solar irradiance, ambient temperature, and wind speed data for 1 year from the installation field are required. The effect of minimum LCOE and maximum annual energy objective functions on the PV plant design was determined. The HGWOSCA optimisation technique and a single SCA algorithm were applied with 400 search agents and 30 iterations to solve the design problem.

According to the results presented in Table 7, the PV plant optimal design variables depend on the selected objective function. The minimum LCOE and maximum annual energy result in two completely different optimal PV plant structures. PV power plant results are presented in

Table 8. The optimisation process applying HGWOSCA outperforms the single SCA for minimum and maximum objective functions.

Table 7. Optimal design variables using semi-hourly measurement data.

Design Variables	Minimum Levelised Cost of Energy (LCOE)		Maximum Energy	
	Sine Cosine Algorithm (SCA)	HGWOSCA	SCA	HGWOSCA
N_s	20	18	21	16
N_p	70	74	49	62
N_r	5	4	1	1
β	15	15	15	15
F_y	1.925	1.542	0.640	0.643
PV_i	PV1	PV3	PV3	PV3
INV_i	INV3	INV3	INV3	INV3
PV_{orien}	1	1	2	2

Table 8. Results of optimal design algorithms using semi-hourly measurement data.

PV Plant Parameters	Minimum LCOE		Maximum Energy	
	SCA	HGWOSCA	SCA	HGWOSCA
LCOE ($/MWh)	29.1829	**28.6283**	32.0983	32.1174
Yearly total energy (MWh)	731.4199	776.4012	785.8698	**786.5035**
Total energy (GWh)	18.2855	19.410	19.6467	19.6626
Total cost (M$)	0.5336	0.5557	0.6305	0.6315
Installation Cost (M$)	0.4465	0.4632	0.5368	0.5378
Maintenance cost (M$)	0.0871	0.0925	0.0936	0.0937
Junction Boxes	6	7	18	18
PV modules (N_I)	1365	1376	1393	1394
Inverters (N_i)	1	1	2	2
Rows	11	14	36	36

For both objectives using HGWOSCA, the optimisation process has selected mono-crystalline PV module type 3 (PV3) from the list of candidates. This module uses 295 W, and inverter type 3 (INV3) was selected from a list of three inverters. This inverter uses 500 kW and presents the central topology of the PV power plant. With the objective of maximum annual energy, the suggested number of PV modules is 1394 and 2 inverter to have 786.5035 (MWh). In this case, the LCOE was 32.1174 ($/MWh), and the PV plant total cost was the highest at 0.6315 (M$). With the objective of minimum LCOE, the number of PV modules is 1376 and only 1 inverter is required to have 28.6283 ($/MWh) of LCOE. In this case, the annual energy generation is equal to 786.5035 (MWh), and the total cost was reduced to 0.5557 (M$) compared with the first case. The use of LCOE's objective function to optimise the design of PV plants can reduce the financial risks, as proven in this case study. The total cost of using minimum LCOE decreased by 12% with a benefit of 71,800 ($) in terms of installation cost, maintenance and operation costs. Figure 9 illustrates the maintenance and operational costs and the installation cost throughout the life of the PV plant for minimum LCOE and maximum annual energy generation.

The area occupied by the PV power plant can be calculated based on the summation of the occupied area by all PV rows, according to the length of each row and the inter-row area of all adjacent rows. The total available area of the installation field is equal to 3131 m^2 and the installed PV modules occupied 3094 m^2 of the installation site, which is nearly the same as the total area of the field. Therefore, the percentage of the occupied area by PV modules in the two cases presents 99% of the available area. The arrangement of PV modules in rows within the installation area is illustrated in Figure 10 using the LCOE objective function. The length of each row changed from one row to another according to the shape of the PV plant. Furthermore, this configuration has been designed in terms of the shape of the installation area, reflecting the actual situation. The difference obtained on the energy production using LCOE and maximum energy objective functions is due to the configuration and the arrangement of the PV modules within the available installation area. On the one hand, the optimal design of the PV plant under the maximum annual energy resulted in the minimum number of lines N_r installed in each row, which is equal to 1. Additionally, this arrangement allowed the PV modules to capture more reflected radiation from the ground. Furthermore, at $N_r = 1$, a small distance between two adjacent rows in terms of shading effect is required, thereby increasing the total number of rows in the installation area to $D_{row} = 36$ with one PV module line in each row and increasing the reflected radiation on PV modules. Moreover, the total number of PV modules for maximum energy is equal to 1394 and distributed among two central inverters. However, the number of PV modules for LCOE is less, leads to 1379 and arranged among only one inverter. PV modules are installed in multiple lines in case of LCOE objective function. In this configuration, the number of lines N_r for each row is equal to 4 and leads only to 14 rows. Moreover, this configuration decreases the reflected radiation from the ground to be captured by PV modules and cannot be absorbed by the rest of the lines ($N_r > 1$). The PV modules are installed horizontally for minimum LCOE ($PV_{orien} = 1$) and vertically ($PV_{orien} = 2$) for maximum annual energy.

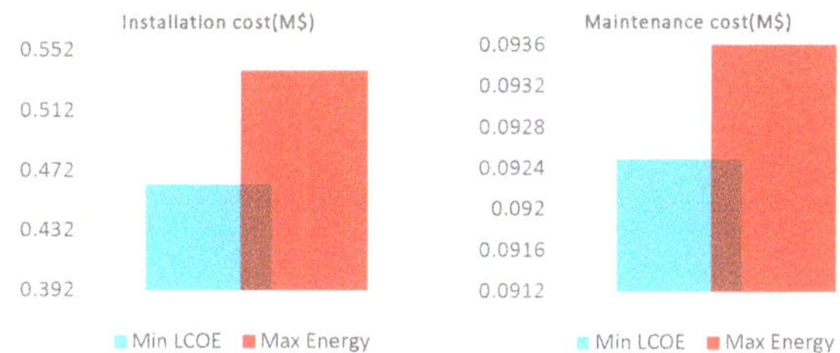

Figure 9. Throughout the life of the PV plant optimised by HGWOSCA.

Figure 10. Rows arrangement for minimum LCOE using HGWOSCA.

Figure 11 illustrates the monthly energy generation by the PV power plant for the LCOE objective function. The PV plant energy generation remained high over the year, with an energy average of 65 (MWh) per month. The highest value of the energy generated by the PV power plant is obtained in March, because this condition is due to the high solar irradiance in this period.

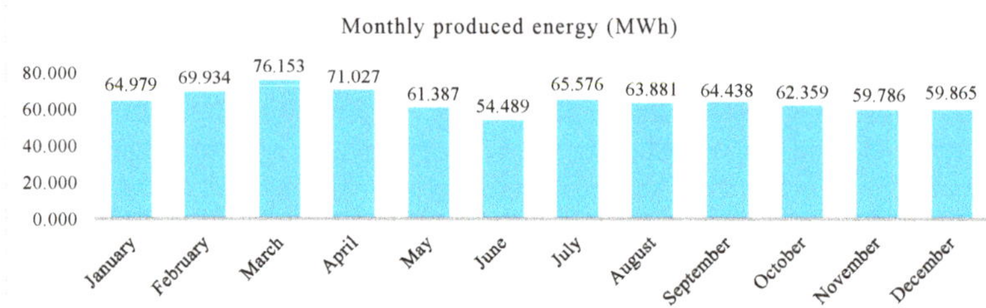

Figure 11. PV plant energy generation (MWh).

For comparison, the semi-hourly average time was compared with the hourly average time meteorological data to examine the step time effect on the PV plant performance. The peaks of the meteorological data can influence the design solution. Therefore, the usage of annual semi-hourly average time rather than monthly, daily and hourly is recommended, as semi-hourly data contain the troughs and peaks of solar irradiation, ambient temperature, and wind speed. According to the results presented in Tables 9 and 10, the step time data can affect the objective functions. The LCOE for semi-hourly average time is 28.6283 ($/MWh), and that obtained for hourly average time is higher and equal to 28.637 ($/MWh). The use of semi-hourly average time meteorological data in designing the PV plant can increase the financial benefits.

Table 9. Optimal design variables using hourly measurement data.

Design Variables	Minimum LCOE		Maximum Energy	
	SCA	**HGWOSCA**	**SCA**	**HGWOSCA**
N_s	19	18	22	14
N_p	70	74	42	47
N_r	5	4	1	1
β	15	15	15	15
F_y	1.925	1.540	0.64	0.64
PV_i	PV3	PV3	PV3	PV3
INV_i	INV3	INV3	INV3	INV3
PV_{orien}	1	1	2	2

Table 10. Results of optimal design algorithms using hourly measurement data.

PV Plant Parameters	Minimum LCOE		Maximum Energy	
	SCA	**HGWOSCA**	**SCA**	**HGWOSCA**
LCOE ($/MWh)	28.6622	**28.6370**	32.0983	32.1552
Yearly total energy (MWh)	770.0032	776.1651	785.6774	**786.2915**
Total energy (GWh)	19.2501	19.4041	19.6419	19.6573
Total cost (M$)	0.5517	0.5557	0.6305	0.6321
Installation Cost (M$)	0.4600	0.4632	0.5368	0.5384
Maintenance cost (M$)	0.0917	0.0925	0.0936	0.0937
Junction Boxes	6	7	18	18
PV modules (N_I)	1365	1376	1393	1394
Inverters (N_i)	1	1	2	2
Rows	11	14	36	36

In all resulting cases, the proposed HGWOSCA optimisation approach was applied successfully and showed higher efficiency than that of a single SCA technique, with high performance in determining the optimal solution and solving the PV plant complex design problem. The convergence optimisation of annual energy and LCOE is illustrated in Figures 12 and 13.

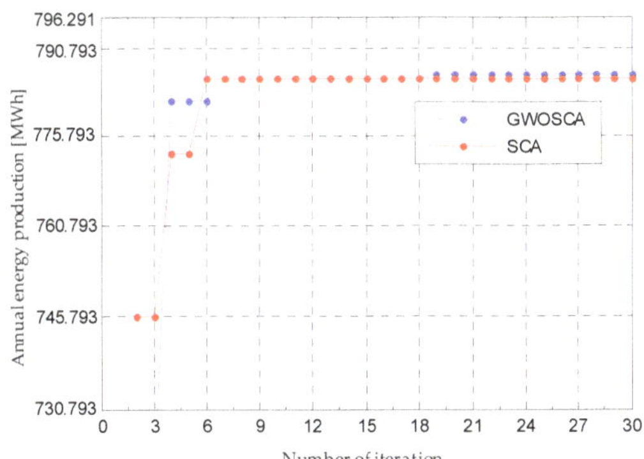

Figure 12. Convergence of the optimisation of annual energy using HGWOSCA algorithm for semi-hourly data.

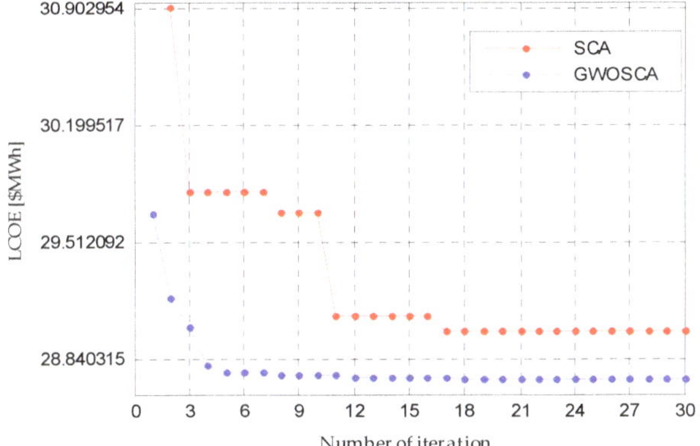

Figure 13. The convergence of the optimisation of LCOE using HGWOSCA algorithm for semi-hourly data.

Effect of PV Module Reduction Coefficient

A sensitivity analysis was applied to evaluate the PV power plant performance. Accordingly, the variations in the PV module annual reduction coefficient were investigated. The optimisation results were obtained for different annual reduction coefficient values, from 0.3% to 0.7% per year. The annual reduction coefficient used in this study was 0.5%, as mentioned in Equation (44).

The optimum results for five different values for the annual reduction coefficient of the PV module are presented in Figures 14 and 15. According to the results, by increasing the PV module reduction coefficient, the PV plant energy production is reduced throughout its lifetime period. The LCOE of the PV plant increases by increasing the PV module reduction coefficient. By contrast, the total cost of the PV power plant is not affected and has the same value for all reduction coefficient values.

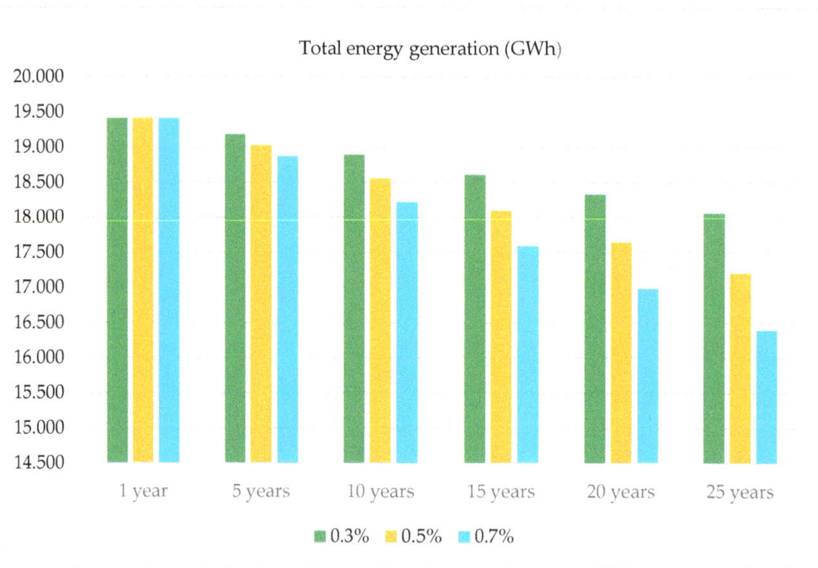

Figure 14. Total energy for reduction coefficient variations.

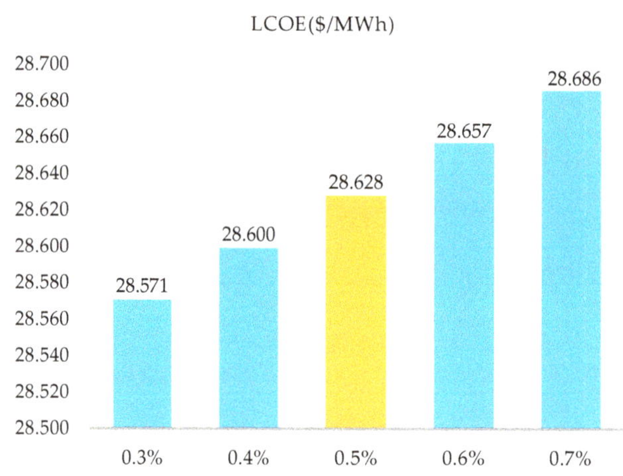

Figure 15. LCOE for reduction coefficient variations.

In economic terms, an improved PV module annual reduction coefficient leads to the recovery of capital investment of the PV plant within a smaller time period, making the PV plant economically profitable. Moreover, the sensitivity of the PV power plant improved by the decrement of the PV module annual reduction coefficient and vice versa.

6. Conclusions

The proposed methodology was executed using semi-hourly time-resolution (i.e., 30 min-average) values of meteorological input data, including solar irradiance, ambient temperature, and wind speed. The procedure considers PV modules and inverter specifications, including a list of different commercially available PV modules and inverter technologies as candidates. The optimisation process selects only one PV module and inverter from a list of several alternatives, presenting the optimum combination. The proposed PV plant area model considers the shape and size of the installation field to properly arrange all the existing components.

The minimum LCOE and maximum annual energy objective functions were used to design the PV power plant. On the basis of the optimal results, the total cost of using the minimum LCOE objective function decreased by 12% with a benefit of 71,800 ($), including installation cost and maintenance and operation costs compared with the maximum annual energy. In this methodology, the HGWOSCA optimisation technique and a single SCA algorithm were applied. The optimum design solution shows that the proposed HGWOSCA is more efficient. Additionally, the PV plant optimal design variables depend on the selected objective function. The minimum LCOE and maximum annual energy result in two different optimal PV plant structures. LCOE improved with the use of semi-hourly average time meteorological data for designing the PV plant and can increase the financial benefits. Moreover, the sensitivity analysis shows that the PV power plant can be improved by the decrement of the PV module annual reduction coefficient and makes the PV plant economically more profitable.

Author Contributions: T.E.K.Z. contributed theoretical approaches, simulation, and preparing the article; M.R.A., M.F.N.T., S.M.Z. and A.D. contributed to supervision; M.R.A., A.D. and S.M. contributed to article editing. All authors have read and agreed to the published version of the manuscript.

Funding: This research was funded by the School of Electrical System Engineering Research Fund (SESERF), UniMAP.

Acknowledgments: We gratefully acknowledge the support of the Algerian company of electricity SONELGAZ, for providing the measurement data.

Conflicts of Interest: The authors declare no conflict of interest.

References

1. Mohammadi, K.; Naderi, M.; Saghafifar, M. Economic feasibility of developing grid-connected photovoltaic plants in the southern coast of Iran. *Energy* **2018**, *156*, 17–31. [CrossRef]
2. Zou, H.; Du, H.; Brown, M.A.; Mao, G. Large-scale PV power generation in China: A grid parity and techno-economic analysis. *Energy* **2017**, *134*, 256–268. [CrossRef]
3. Feldman, D.; Margolis, R. *Q2/Q3 2018 Solar Industry Update 2018*; Technical Report; National Renewable Energy Lab. (NREL): Golden, CO, USA, 2018.
4. Comello, S.; Reichelstein, S.; Sahoo, A. The road ahead for solar PV power. *Renew. Sustain. Energy Rev.* **2018**, *92*, 744–756. [CrossRef]
5. Mondol, J.D.; Yohanis, Y.G.; Norton, B. Optimal sizing of array and inverter for grid-connected photovoltaic systems. *Sol. Energy* **2006**, *80*, 1517–1539. [CrossRef]
6. Kornelakis, A.; Marinakis, Y. Contribution for optimal sizing of grid-connected PV-systems using PSO. *Renew. Energy* **2010**, *35*, 1333–1341. [CrossRef]
7. Notton, G.; Lazarov, V.; Stoyanov, L. Optimal sizing of a grid-connected PV system for various PV module technologies and inclinations, inverter efficiency characteristics and locations. *Renew. Energy* **2010**, *35*, 541–554. [CrossRef]
8. Sulaiman, S.I.; Rahman, T.K.A.; Musirin, I.; Shaari, S.; Sopian, K. An intelligent method for sizing optimization in grid-connected photovoltaic system. *Sol. Energy* **2012**, *86*, 2067–2082. [CrossRef]
9. Chen, S.; Li, P.; Brady, D.; Lehman, B. Determining the optimum grid-connected photovoltaic inverter size. *Sol. Energy* **2013**, *87*, 96–116. [CrossRef]
10. Perez-Gallardo, J.R.; Azzaro-Pantel, C.; Astier, S.; Domenech, S.; Aguilar-Lasserre, A. Ecodesign of photovoltaic grid-connected systems. *Renew. Energy* **2014**, *64*, 82–97. [CrossRef]
11. Senol, M.; Abbasoğlu, S.; Kukrer, O.; Babatunde, A. A guide in installing large-scale PV power plant for self consumption mechanism. *Sol. Energy* **2016**, *132*, 518–537. [CrossRef]
12. Báez-Fernández, H.; Ramirez-Beltran, N.D.; Méndez-Piñero, M.I. Selection and configuration of inverters and modules for a photovoltaic system to minimize costs. *Renew. Sustain. Energy Rev.* **2016**, *58*, 16–22. [CrossRef]

13. Topić, D.; Knežević, G.; Fekete, K. The mathematical model for finding an optimal PV system configuration for the given installation area providing a maximal lifetime profit. *Sol. Energy* **2017**, *144*, 750–757. [CrossRef]

14. Durusu, A.; Erduman, A. An improved methodology to design large-scale photovoltaic power plant. *J. Sol. Energy Eng.* **2017**, *140*, 011007. [CrossRef]

15. Zidane, T.E.K.; Adzman, M.R.; Zali, S.M.; Mekhilef, S.; Durusu, A.; Tajuddin, M.F.N. Cost-effective topology for photovoltaic power plants using optimization design. In Proceedings of the 2019 IEEE 7th Conference on Systems, Process and Control (ICSPC), Malacca, Malaysia, 13–14 December 2019; pp. 210–215.

16. Wang, H.; Muñoz-García, M.; Moreda, G.; Alonso-Garcia, C. Optimum inverter sizing of grid-connected photovoltaic systems based on energetic and economic considerations. *Renew. Energy* **2018**, *118*, 709–717. [CrossRef]

17. Perez-Gallardo, J.R.; Azzaro-Pantel, C.; Astier, S. Combining multi-objective optimization, principal component analysis and multiple criteria decision making for ecodesign of photovoltaic grid-connected systems. *Sustain. Energy Technol. Assess.* **2018**, *27*, 94–101. [CrossRef]

18. Zidane, T.E.K.; Adzman, M.R.; Tajuddin, M.F.N.; Zali, S.M.; Durusu, A. Optimal configuration of photovoltaic power plant using grey wolf optimizer: A comparative analysis considering CdTe and c-Si PV modules. *Sol. Energy* **2019**, *188*, 247–257. [CrossRef]

19. Bakhshi-Jafarabadi, R.; Sadeh, J.; Soheili, A. Global optimum economic designing of grid-connected photovoltaic systems with multiple inverters using binary linear programming. *Sol. Energy* **2019**, *183*, 842–850. [CrossRef]

20. Sulaiman, S.I.; Rahman, T.K.A.; Musirin, I.; Shaari, S. Sizing grid-connected photovoltaic system using genetic algorithm. In Proceedings of the 2011 IEEE Symposium on Industrial Electronics and Applications, Langkawi, Malaysia, 25–28 September 2011; pp. 505–509. [CrossRef]

21. Al-Sabounchi, A.M.; Yalyali, S.A.; Al-Thani, H.A. Design and performance evaluation of a photovoltaic grid-connected system in hot weather conditions. *Renew. Energy* **2013**, *53*, 71–78. [CrossRef]

22. Kornelakis, A.; Koutroulis, E. Methodology for the design optimisation and the economic analysis of grid-connected photovoltaic systems. *IET Renew. Power Gener.* **2009**, *3*, 476. [CrossRef]

23. Koutroulis, E.; Yang, Y.; Blaabjerg, F. Co-Design of the PV array and DC/AC inverter for maximizing the energy production in grid-connected applications. *IEEE Trans. Energy Convers.* **2018**, *34*, 509–519. [CrossRef]

24. Moradi-Shahrbabak, Z.; Tabesh, A.; Yousefi, G.R. Economical design of utility-scale photovoltaic power plants with optimum availability. *IEEE Trans. Ind. Electron.* **2013**, *61*, 3399–3406. [CrossRef]

25. Okido, S.; Takeda, A. Economic and environmental analysis of photovoltaic energy systems via robust optimization. *Energy Syst.* **2013**, *4*, 239–266. [CrossRef]

26. Paravalos, C.; Koutroulis, E.; Samoladas, V.; Kerekes, T.; Sera, D.; Teodorescu, R. Optimal design of photovoltaic systems using high time-resolution meteorological data. *IEEE Trans. Ind. Inform.* **2014**, *10*, 2270–2279. [CrossRef]

27. Ramli, M.; Hiendro, A.; Sedraoui, K.; Twaha, S. Optimal sizing of grid-connected photovoltaic energy system in Saudi Arabia. *Renew. Energy* **2015**, *75*, 489–495. [CrossRef]

28. Rodrigo, P.M. Improving the profitability of grid-connected photovoltaic systems by sizing optimization. In Proceedings of the 2017 IEEE Mexican Humanitarian Technology Conference (MHTC), Puebla, Mexico, 29–31 March 2017; pp. 1–6.

29. Arefifar, S.A.; Paz, F.; Ordonez, M. Improving solar power PV plants using multivariate design optimization. *IEEE J. Emerg. Sel. Top. Power Electron.* **2017**, *5*, 638–650. [CrossRef]

30. Camps, X.; Velasco, G.; De La Hoz, J.; Martín, H. Contribution to the PV-to-inverter sizing ratio determination using a custom flexible experimental setup. *Appl. Energy* **2015**, *149*, 35–45. [CrossRef]

31. Demoulias, C. A new simple analytical method for calculating the optimum inverter size in grid-connected PV plants. *Electr. Power Syst. Res.* **2010**, *80*, 1197–1204. [CrossRef]

32. Fernández-Infantes, A.; Contreras, J.; Bernal-Agustín, J.L. Design of grid connected PV systems considering electrical, economical and environmental aspects: A practical case. *Renew. Energy* **2006**, *31*, 2042–2062. [CrossRef]

33. Gemignani, M.; Rostegui, G.; Salles, M.B.C.; Kagan, N. Methodology for the sizing of a solar PV plant: A comparative case. In Proceedings of the 2017 6th International Conference on Clean Electrical Power (ICCEP), Santa Margherita Ligure, Italy, 27–29 June 2017; pp. 7–13.

34. Kerekes, T.; Koutroulis, E.; Sera, D.; Teodorescu, R.; Katsanevakis, M. An optimization method for designing large PV Plants. *IEEE J. Photovoltaics* **2012**, *3*, 814–822. [CrossRef]

35. Kerekes, T.; Koutroulis, E.; Eyigun, S.; Teodorescu, R.; Katsanevakis, M.; Sera, D. A practical optimization method for designing large PV plants. In Proceedings of the 2011 IEEE International Symposium on Industrial Electronics, Gdansk, Poland, 27–30 June 2011; pp. 2051–2056. [CrossRef]

36. Kornelakis, A. Multiobjective particle swarm optimization for the optimal design of photovoltaic grid-connected systems. *Sol. Energy* **2010**, *84*, 2022–2033. [CrossRef]

37. Alsadi, S.; Khatib, T. Photovoltaic power systems optimization research status: A review of criteria, constrains, models, techniques, and software tools. *Appl. Sci.* **2018**, *8*, 1761. [CrossRef]

38. Baños, R.; Manzano-Agugliaro, F.; Montoya, F.; Gil, C.; Alcayde, A.; Gómez, J. Optimization methods applied to renewable and sustainable energy: A review. *Renew. Sustain. Energy Rev.* **2011**, *15*, 1753–1766. [CrossRef]

39. Eltawil, M.A.; Zhao, Z. Grid-connected photovoltaic power systems: Technical and potential problems—A review. *Renew. Sustain. Energy Rev.* **2010**, *14*, 112–129. [CrossRef]

40. Khatib, T.; Mohamed, A.; Sopian, K. A review of photovoltaic systems size optimization techniques. *Renew. Sustain. Energy Rev.* **2013**, *22*, 454–465. [CrossRef]

41. Mellit, A.; Kalogirou, S.; Hontoria, L.; Shaari, S. Artificial intelligence techniques for sizing photovoltaic systems: A review. *Renew. Sustain. Energy Rev.* **2009**, *13*, 406–419. [CrossRef]

42. Maatallah, T.; El Alimi, S.; Ben Nassrallah, S. Performance modeling and investigation of fixed, single and dual-axis tracking photovoltaic panel in Monastir city, Tunisia. *Renew. Sustain. Energy Rev.* **2011**, *15*, 4053–4066. [CrossRef]

43. Singh, N.; Singh, S. A novel hybrid GWO-SCA approach for optimization problems. *Eng. Sci. Technol. Int. J.* **2017**, *20*, 1586–1601. [CrossRef]

44. Mirjalili, S. SCA: A sine cosine algorithm for solving optimization problems. *Knowl.-Based Syst.* **2016**, *96*, 120–133. [CrossRef]

45. Energy Information Administration (EIA) USA. Total Electricity Net Generation. 2014. Available online: https://www.eia.gov/beta/international/data/browser (accessed on 30 March 2015).

46. Fodhil, F.; Hamidat, A.; Nadjemi, O. Potential, optimization and sensitivity analysis of photovoltaic-diesel-battery hybrid energy system for rural electrification in Algeria. *Energy* **2019**, *169*, 613–624. [CrossRef]

47. Saiah, S.B.D.; Stambouli, A.B. Prospective analysis for a long-term optimal energy mix planning in Algeria: Towards high electricity generation security in 2062. *Renew. Sustain. Energy Rev.* **2017**, *73*, 26–43. [CrossRef]

48. Alegria CDER. 2010. Available online: http://www.cder.dz (accessed on 12 April 2010).

49. Abada, Z.; Bouharkat, M. Study of management strategy of energy resources in Algeria. *Energy Rep.* **2018**, *4*, 1–7. [CrossRef]

50. Solar Resource Maps and GIS Data for 200+ Countries. Available online: https://solargis.com/maps-and-gis-data/download/algerian.d. (accessed on 7 June 2017).

51. Stambouli, A.B. Promotion of renewable energies in Algeria: Strategies and perspectives. *Renew. Sustain. Energy Rev.* **2011**, *15*, 1169–1181. [CrossRef]

52. World Bank. World Development Indicators. 2017. Available online: http://data.worldbank.org/data-catalog/world-development-indicators (accessed on 1 March 2017).

53. Bouznit, M.; Pablo-Romero, M.D.P. CO_2 emission and economic growth in Algeria. *Energy Policy* **2016**, *96*, 93–104. [CrossRef]

54. Boukhelkhal, A.; Bengana, I. Cointegration and causality among electricity consumption, economic, climatic and environmental factors: Evidence from North-Africa region. *Energy* **2018**, *163*, 1193–1206. [CrossRef]

55. Laib, I.; Hamidat, A.; Haddadi, M.; Ramzan, N.; Olabi, A. Study and simulation of the energy performances of a grid-connected PV system supplying a residential house in north of Algeria. *Energy* **2018**, *152*, 445–454. [CrossRef]

56. Mihoub, S.; Chermiti, A.; Beltagy, H. Methodology of determining the optimum performances of future concentrating solar thermal power plants in Algeria. *Energy* **2017**, *122*, 801–810. [CrossRef]

57. Ministry of Energy Algeria. Algerian Energy Review N4. May 2015; ISSN 2437-0479. Available online: www.mem-algeria.org (accessed on 15 July 2015).

58. Ministry of Energy Algeria. Objectives New Renewable Energies Program in Algeria (2015-2020-2030). 2015. Available online: www.mem-algeria.org (accessed on 5 September 2015).

59. The Weekend Read: Sustainable Finance—Reassessing Risk, Purpose. Available online: https://www.pv-magazine.com/2019/08/16/algerian-diesel-solar-tender-concludes-with-lowest-bid-of-11-cents/n.d. (accessed on 16 August 2019).

60. Fadaee, M.; Radzi, M.A.M. Multi-objective optimization of a stand-alone hybrid renewable energy system by using evolutionary algorithms: A review. *Renew. Sustain. Energy Rev.* **2012**, *16*, 3364–3369. [CrossRef]

61. Panchula, A.F.; Hayes, W.; Bilash, J.; Kimber, A.; Graham, C. First year performance of a 20 MWac PV power plant. In Proceedings of the 2011 37th IEEE Photovoltaic Specialists Conference, Washington, DC, USA, 19–24 June 2011; p. 001993.

62. Krishnamoorthy, H.S.; Essakiappan, S.; Enjeti, P.N.; Balog, R.; Ahmed, S. A new multilevel converter for Megawatt scale solar photovoltaic utility integration. In Proceedings of the 2012 Twenty-Seventh Annual IEEE Applied Power Electronics Conference and Exposition (APEC), Orlando, FL, USA, 5–9 February 2012; pp. 1431–1438.

63. Marcos, J.; De La Parra, I.; García, M.; Marroyo, L.; Cirés, E. The potential of forecasting in reducing the LCOE in PV plants under ramp-rate restrictions. *Energy* **2019**, *188*, 116053. [CrossRef]

64. Sharma, V.; Kumar, A.; Sastry, O.S.; Chandel, S. Performance assessment of different solar photovoltaic technologies under similar outdoor conditions. *Energy* **2013**, *58*, 511–518. [CrossRef]

65. Cabrera-Tobar, A.; Bullich-Massagué, E.; Aragüés-Peñalba, M.; Gomis-Bellmunt, O. Topologies for large scale photovoltaic power plants. *Renew. Sustain. Energy Rev.* **2016**, *59*, 309–319. [CrossRef]

66. Testa, A.; De Caro, S.; La Torre, R.; Scimone, T. A probabilistic approach to size step-up transformers for grid connected PV plants. *Renew. Energy* **2012**, *48*, 42–51. [CrossRef]

67. Noorian, A.M.; Moradi, I.; Kamali, G.A. Evaluation of 12 models to estimate hourly diffuse irradiation on inclined surfaces. *Renew. Energy* **2008**, *33*, 1406–1412. [CrossRef]

68. Demain, C.; Journée, M.; Bertrand, C. Evaluation of different models to estimate the global solar radiation on inclined surfaces. *Renew. Energy* **2013**, *50*, 710–721. [CrossRef]

69. Liu, B.Y.H.; Jordan, R.C. A rational procedure for predicting the long-term average performance of flat-plate solar-energy collectors with design data for the U. S., its outlying possessions and Canada. *Sol. Energy* **1963**, *7*, 53–74. [CrossRef]

70. Ayaz, R.; Durusu, A.; Akca, H. Determination of optimum tilt angle for different photovoltaic technologies considering ambient conditions: A case study for Burdur, Turkey. *J. Sol. Energy Eng.* **2017**, *139*, 041001. [CrossRef]

71. Chin, V.J.; Chin, V.J.; Ishaque, K. Cell modelling and model parameters estimation techniques for photovoltaic simulator application: A review. *Appl. Energy* **2015**, *154*, 500–519. [CrossRef]

72. Orioli, A.; Di Gangi, A. A procedure to evaluate the seven parameters of the two-diode model for photovoltaic modules. *Renew. Energy* **2019**, *139*, 582–599. [CrossRef]

73. Elbaset, A.A.; Ali, H.; Sattar, M.A.-E.; Elbaset, A.A. Novel seven-parameter model for photovoltaic modules. *Sol. Energy Mater. Sol. Cells* **2014**, *130*, 442–455. [CrossRef]

74. Dhoke, A.; Sharma, R.; Saha, T.K. PV module degradation analysis and impact on settings of overcurrent protection devices. *Sol. Energy* **2018**, *160*, 360–367. [CrossRef]

75. Rosillo, F.G.; Alonso-Garcia, C. Evaluation of color changes in PV modules using reflectance measurements. *Sol. Energy* **2019**, *177*, 531–537. [CrossRef]

76. Deline, C.; Deline, C. Partially shaded operation of a grid-tied PV system. In Proceedings of the 2009 34th IEEE Photovoltaic Specialists Conference (PVSC), Philadelphia, PA, USA, 7–12 June 2009; pp. 001268–001273. [CrossRef]

77. Mirjalili, S.; Mirjalili, S.M.; Lewis, A. Grey Wolf Optimizer. *Adv. Eng. Softw.* **2014**, *69*, 46–61. [CrossRef]

Article

A Coordinated Voltage and Reactive Power Control Architecture for Large PV Power Plants

Massimiliano Chiandone, Riccardo Campaner, Daniele Bosich * and Giorgio Sulligoi

Department of Engineering and Architecture, University of Trieste, 34127 Trieste, Italy;
mchiandone@units.it (M.C.); riccardo.campaner@gmail.com (R.C.); gsulligoi@units.it (G.S.)
* Correspondence: dbosich@units.it

Received: 11 April 2020; Accepted: 9 May 2020; Published: 13 May 2020

Abstract: The increasing presence of nonprogrammable renewable energy sources (RES) forces towards the development of new methods for voltage control. In the case of centralized generation, the hierarchical regulation or secondary voltage regulation (SVR) is guaranteed by coordinated voltage and reactive power controls in transmission systems. This type of regulation loses effectiveness when the generation becomes distributed and based on small and medium sized generators. To overcome this problem, it is important that also distributed generators, typically based on RES, participate in the voltage regulation. By starting from the methodologies already applied, this work wants to present a new method for involving distributed generators in SVR. The novelty is given by the application of an existing methodology to the new configuration of electrical grids characterized by a relevant distributed generation. The aim is to control the distributed generators (DGs) as coordinated sources of reactive power for conveniently supporting the voltage regulation. In this paper, a real large photovoltaic (PV) plant is considered. The power plant is composed of several PV generators connected through a distribution network. With the algorithm proposed, the set of generators can be treated as a single traditional power plant that can participate in the hierarchical voltage regulation. The reactive power of each single generator is coordinated in a way similar to the SVR used in several national systems.

Keywords: distribution network; hierarchical voltage control; reactive power control; RES; SVR; PV

---

1. Introduction

As well known in electrical engineering, user appliances work in the best conditions (i.e., performance, efficiency, and lifetime) when fed at a rated voltage or within a small voltage deviation from that value [1]. Not only the loads' section but similar considerations can also be drawn for production, transmission, and distribution systems (e.g., generators, transformers, lines, reactors, and shunt capacitors). Indeed, also in these cases, the voltage on components is to be maintained within a limited range for avoiding various negative effects on the system operation [2,3].

During a standard operative scenario in transmission grids, undesired voltage fluctuations at the grid nodes are mainly caused by variations in absorbed power, which is variously requested from the different loads connected to the network [4]. On the other hand, temporary out of service of any network component (lines, transformers, etc.) are responsible for significant variations in supplied voltage and even the loads' disconnection [4]. For the reasons expressed so far, the transmission system operator (TSO) must guarantee an adequate voltage control service, i.e., a complex of measures for achieving a suitable voltage control in the different network nodes [4]. Clearly, each TSO operating on a specific grid can implement a peculiar voltage control service, which is definitely not stiffly established. Being the high voltage (HV) nodes mainly influenced by the flows of reactive power [4], the voltage control is typically performed by regulating the reactive power flows by means of different reactive power resources. For giving some examples, it possible to mention FACT devices, synchronous generators, synchronous condenser, transformer tap changers, static VAR compensators, capacitor banks, and capacitance of lines and cables [5].

During the last decades, several countries have implemented hierarchical voltage control systems, practically based on the reactive power provided by conventional large power plants [6]. In this regard, a review of the adopted systems is reported in [5]. As a matter of fact, the standard voltage control systems have been designed and developed for traditional electric power grids, which are characterized by a unidirectional energy flow (from large production centers up to the loads). Nowadays, this assumption is not valid anymore, where the electric power systems are experimenting a great structural change by moving from the centralized production paradigm to the distributed generation model [7,8]. The latter is characterized by small–medium size generation centers, typically based on renewable energy sources (RES), which are conveniently (sometime randomly) distributed throughout the territory. The large penetration of RES leads relevant consequences on several aspects (power quality, power losses, and voltage profiles), as discussed in [9–11].

The RES scenario in EU-28 is constantly evolving [12]. Just to give some data [12], at the end of 2014, the installed RES power has been equal to 369,511 GW, more than a third of total installed power (37.8%). In detail, RES power plants are capable of contributing to the production of 930 TWh, another time almost one-third of the total production (29.2%). This data can be summarized not only in the significant growth rate of RES in the last years but also in the fact that RES growth is the only one having a positive index among the different resources. There are several technologies that contribute to renewable generation: among others, hydroelectric, wind, biomass, and photovoltaic [12–14]. Particularly, the last references provide some important reports for the context frame.

Nowadays, the secondary voltage regulation (SVR) for HV networks is based on traditional fossil power plants and large hydroelectric power plants, while most of the new renewable energy plants are not taking part in this task [5]. By considering this aspect, the existing voltage control systems are experiencing a decrease in the control capability, especially when the power of conventional thermal plants is replaced by the one produced by distributed generators (DGs) [5]. For what concerns the connection, medium voltage (MV) and low voltage (LV) distribution networks are the standard end-point for RES power plants, whilst the HV transmission system represents the future connecting point for high-power renewable sources. In this regard, the Italian case represents an interesting example, where 6.4% of PV plants are directly connected to the HV grid by the end of 2014 [15]. The Italian case is noteworthy also for the high presence of nonprogrammable RES power plants, whereas the Italian TSO implements a hierarchical voltage control by adopting the production plants as actuators [5].

Independently from the voltage level of interconnection, the introduction of RES technology can determine important consequences on the voltage control [16]. The RES operation outside the constant unitary power factor is commonly accepted as a standard requirement, independently from the interconnection at transmission or distribution level [17–20]. Particularly, last references are aimed at defining regulations and guidelines for network code.

In this regard, the authors have largely studied the contribution of photovoltaic (PV) plants in supporting the network voltage: in [21], the voltage control functionality is guaranteed by modulating the reactive power and bounding the injected ramp of active power, while PV power plants are regulated for behaving as STATCOM devices in [22]. As proposed in [23], PV units can provide reactive power compensation as ancillary service, whilst the dynamic reactive power compensation can be obtained by suitably controlling power electronics devices [24]. By regulating active and reactive power of PV systems, the adaptive droop-based control algorithms are useful to minimize losses and increase the capacity installation, while avoiding overvoltage [25]. For what concerns the integration of PV systems into a residential area, the papers [26] and [27] offer some interesting proposals, while the mix of RES and electrical storage is conversely described in [21,28–30]. Extending the concept of RES away from the PV power plants, several voltage controls have been proposed in literature [31–35] for the wind farms. Particularly, [36] provides important considerations on how to regulate the wind generators for emulating a hierarchical voltage control. A similar perspective is the one provided by the so-called virtual power plants (VPP) [37]. In such a case, multiple distributed generation plants are connected and managed for behaving as a virtual smart network (VSN). For the electrical grid point of view, the VSN constitutes a single provider of main services (energy and capacity) and auxiliary ones (regulations, reserves, etc.) [38–41].

The present paper wants to analyze an innovative control strategy to conveniently integrate RES power plants (i.e., large PV) into the voltage control system of the transmission network. Such a strategy is based on a hierarchical control architecture, successfully implemented in several countries [6,42–44]. A crucial device for this control system is the reactive power regulator (RPR), which is implemented in the so-called SART (in Italian language "Sistema Automatico per la Regolazione della Tensione", translated as "Automatic Voltage and Reactive Power Regulator") [45] in the Italian Grid Code. In this paper, the algorithm adopted by the control system is extended to coordinate the reactive powers of all generators that participate in the voltage control, despite their size and position in the network (and therefore applicable also to distribution networks). The final effect is the control of the voltage of the bus connecting the power plant to the rest of the system, through the reactive power of the generators. By comparing simulations and data from measurement campaigns on conventional operating power plants, the paper is initially aimed at validating the proposed methodology. Once the methodology is proved on standard cases, the control strategy can be proficiently transferred to the PV test case.

The paper is organized as follows. Section 2 describes the proposed architecture for the voltage control, while Section 3 is focused on modeling aspects, particularly on the differences in representing RES (where large PV power plant is a particular case) and traditional power plants. Then Section 4 discusses the main topic. Once demonstrated, the validity of the proposed methodology by comparing simulation results and measurements data from traditional power plants, such a methodology will be proficiently extended to the PV case. Finally, Section 5 provides a discussion about the obtained results, while Section 6 outlines the conclusions.

2. Hierarchical Voltage Control Architecture

In this paper, the voltage control strategy is based on a hierarchical architecture (Figure 1), where the main controller has been already proposed and utilized in several countries [46,47]. Therefore, the proposed strategy is based on an already implemented algorithm. Indeed, in the last 30 years this algorithm has proved its simple implementation, while at the same time it is scalable to transmission networks of different size and topology. For all these reasons, the application of such an algorithm has been extended to HV networks populated by nonprogrammable energy resources. The proposed architecture presents an external loop and a cluster of internal controllers, where the RPR behaves as a central control unit for coordinating the reactive power from each generator. Such a regulation is therefore adopted to obtain the voltage control on a peculiar pilot bus, named point of connection (POC). By implementing this control structure, the so-obtained reactive power regulation is actually similar to what is achieved by the SVR on transmission grids [48].

The control structure is explained in the scheme of Figure 1. Particularly, the network operator forwards the voltage reference to the control system, while the POC of the RES generation plant is operating as a pilot bus (as defined in SVR). In traditional fossil fuel power plants, all the generators are normally connected through a busbar, while in a more general case all generators are connected through any kind of grid, thus without any topology regularity. In the proposed architecture, the busbar voltage regulator (BVR) plays a crucial role, being responsible for the pilot bus voltage control (time constant T_b near ten seconds). Based on a classical PI (proportional-integral) functionality, the BVR can impose a level of reactive power q_{liv} (between −1 and +1) to be applied by each generator [48] for getting the requested voltage regulation. For particular cases in which the TSO implements a remote regional voltage control architecture, the TSO excludes the BVR function by directly sending an external q_{liv} reference signal to the reactive power control loop (i.e., switch in Figure 1). In both cases, the so-obtained q_{liv} signal is the input for the RPR, where the q_{liv} is multiplied by each generator reactive power limit to define the vector of reactive power references $\overline{Q_{ref}}$. The vector components are then compared to the actual reactive power $\overline{Q}$ values of each generator, thus determining a vector of errors $\overline{\Delta Q}$. The latter is therefore multiplied by the dynamic decoupling (DD) matrix, whose outputs constitute the inputs for the generator reactive power regulators (GRPRs) [49], thus obtaining the reactive power control loop (i.e., the time constant T_Q is approximately few seconds). The reason for adopting the DD matrix is demonstrated by observing the MIMO (multiple-input-multiple-output) characteristic, which is typical of the generator reactive power control loop (i.e., several PI regulators, one for each generator). Instead, the dynamic decoupling matrix is capable of compensating for the mutual interactions, thus decoupling the MIMO reactive power control loop and consequently simplifying the control system design. Thus, the DD application allows the MIMO system to broken down into n single-input-single-output (SISO) loops, where the n generators are modeled by the same transfer function [49].

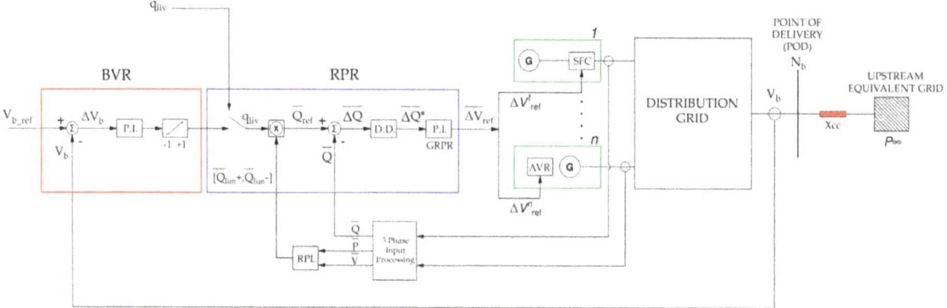

Figure 1. Synoptic scheme of secondary voltage control applied to a general case where all generators are connected by means of a distribution grid with a given topology.

The control signals calculated by GRPRs are the references $\Delta \overline{V_{ref}}$ for each generator, which is represented as static frequency converter SFC and synchronous generators AVR in Figure 2. For what regards the control functionality ensured by BVR and RPR blocks, the related regulator parameters are to be set not only for decoupling internal and external control cycle but also for ensuring a voltage time response with an equivalent time constant of about 50 s. This value is chosen similar to what is usually required in conventional HV production power plants [50]. DGs are thus modeled as "voltage actuators" in terms of a first order mathematical model in d and q-axis coordinates. The time constant T_v of voltage control loops is fast enough compared to that one of the outer reactive power loop T_Q (i.e., under the second), so ΔV_i can be assumed equal to ΔV_{i_ref}. This model can be used for suitably studying the transient stability of the proposed hierarchical voltage control coupled with a simplified RES generator model (Figure 2) [51]. In the past, several studies have investigated the possibility of controlling the reactive power of a voltage source converter (VSC) independently from the active power. For instance, [52] not only shows the reactive power transient response in the presence of a changing in reactive power reference but also points out the P and Q injection decoupling. On the other hand, [53] exhibits a fast response for the reactive power control. This control algorithm can be applied on generation plants with different production technologies, coordinated by the same TSO. The application of the same control scheme allows a sort of uniformity in the dynamic responses of all generators. This control algorithm can be adopted in the case of generation plants with different production technologies, coordinated by the same TSO. Some examples are shown in [54–56], a cluster of hydropower plants, wind, and PV farms.

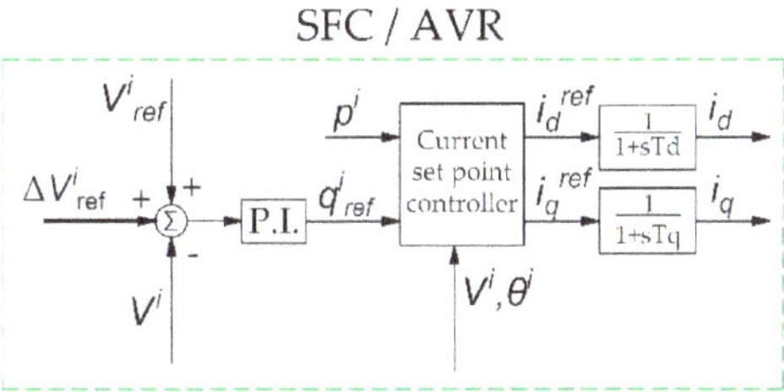

Figure 2. Distributed Generator model.

3. Power Plants Modeling

By observing Figure 1, the importance of the DD matrix appears undeniable, being capable of subdividing the initial system into n independent SISO subsystems. In this regard, the calculation of DD matrix is firstly shown in Section 3.1 for a generic distribution grid connecting all the generators (suitable for distributed RES production power plants). Then, the matrix is provided in Section 3.2 for the standard case, where large traditional generators are connected to the HV busbar through their step-up transformers (i.e., as in traditional large fossil fuel power plants).

3.1. Large RES Production Power Plants

As represented in Figure 3, the large RES production power plants are based on multiple small–medium generators, usually grouped as a medium voltage (MV) cluster interconnected to the single HV POC. This topology is common in practice, regardless of RES source (PV, wind, hydro, etc.) and generators technology (static, synchronous, asynchronous, etc.).

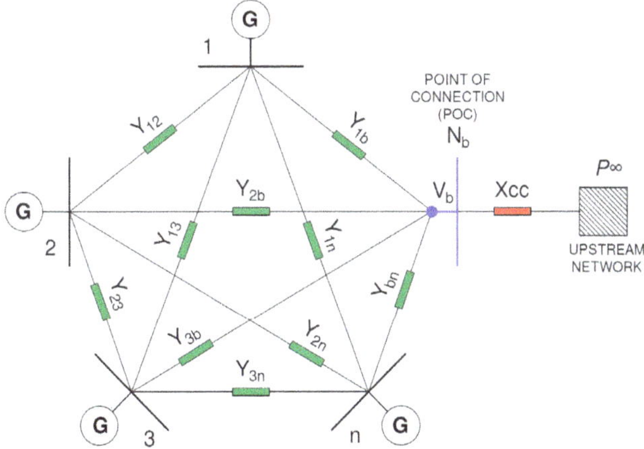

Figure 3. The topology of a renewable energy sources (RES) power plant composed of a medium voltage (MV) network and a unique point of connection (POC) to the high voltage (HV) network.

By adopting the well-known power flow equations, the MV distribution grid can be described by Equation (1), when considering active/reactive power for the node k:

$$\begin{cases} P_k = \sum_{i=1}^{n} V_k \cdot V_i \cdot Y_{ki} \cdot \cos(\theta_k - \theta_i - \gamma_{ki}) \\ Q_k = \sum_{i=1}^{n} V_k \cdot V_i \cdot Y_{ki} \cdot \sin(\theta_k - \theta_i - \gamma_{ki}) \end{cases} \tag{1}$$

where P_k and Q_k are the active/reactive power at the node k, while V_k and V_i are the RMS voltage, respectively, at the node k and node i. The magnitude of admittance coefficients in branch i_k is represented by Y_{ki}, whereas θ_k and θ_i are the voltage phase angle at node k and node i. Finally, γ_{ki} is the admittance phase of branch i_k, whereas n the total number of nodes constituting the analyzed grid. By considering the per-unit notation (i.e., rated values are the basis) and linearizing (1) at a given operating point, the Jacobian matrix is defined as in Equation (2):

$$
\begin{bmatrix} [\Delta p] \\ \\ [\Delta q] \end{bmatrix} = \begin{bmatrix} \left[\frac{dp}{d\theta}\right] & \vline & \left[\frac{dp}{dv}\right] \\ - & & - \\ \left[\frac{dq}{d\theta}\right] & \vline & \left[\frac{dq}{dv}\right] \end{bmatrix} \cdot \begin{bmatrix} [\Delta\theta] \\ \\ [\Delta v] \end{bmatrix}
\tag{2}
$$

By taking into account a behavior around the operating point, the partial derivatives matrices $\left[\frac{dp}{dv}\right], \left[\frac{dp}{d\vartheta}\right], \left[\frac{dq}{dv}\right], \left[\frac{dq}{d\vartheta}\right]$ are the link between active/reactive power and magnitude/phase angle of voltage at buses. As a matter of fact, such matrix coefficients embed the information about the characteristic parameters of the network lines. For the purposes of voltage control, the last equation is to be particularized as in Equation (3), neglecting the active power variations ($[\Delta p] = 0$) and assuming the only reactive power sources as actuators [57]:

$$
[\Delta q] = \left[\left[\frac{dq}{dv}\right] - \left[\frac{dq}{d\theta}\right]\cdot\left[\frac{dp}{d\theta}\right]^{-1}\cdot\left[\frac{dp}{dv}\right]\right][\Delta v]
\tag{3}
$$

By setting a system-operating point, the power flow problem is solved, thus deducing the following equations for the linearized system:

$$
[\Delta q] = [A]\cdot[\Delta v]
\tag{4}
$$

where $[\Delta q]$ and $[\Delta v]$ are the vectors $(n,1)$ of reactive power/voltage variations, whilst the (n,n) matrix $[A]$ models the electric coupling between reactive powers and voltage magnitudes. Hence, the generators are electrically coupled according to the coefficients (5):

$$
\left[\frac{dq}{dv}\right] - \left[\frac{dq}{d\theta}\right]\cdot\left[\frac{dp}{d\theta}\right]^{-1}\cdot\left[\frac{dp}{dv}\right]
\tag{5}
$$

In other words, a voltage variation at every network node causes a reactive power variation in all the n nodes, according to the matrix $[A]$ coefficients, as expressed in Equation (6):

$$
\begin{bmatrix} \Delta q_1 \\ \Delta q_i \\ \Delta q_n \end{bmatrix} = \begin{bmatrix} a_{11} & a_{1i} & a_{1n} \\ a_{i1} & a_{ii} & a_{in} \\ a_{n1} & a_{ni} & a_{nn} \end{bmatrix} \cdot \begin{bmatrix} \Delta v_1 \\ \Delta v_i \\ \Delta v_n \end{bmatrix}
\tag{6}
$$

It is remarkable to notice that $[A]$ is considered full rank in the most practical applications, while the discussion of idiosyncratic cases (i.e., $[A]$ singular) is beyond the study aims. Finally, Equation (7) is capable of modeling the voltage at the POC, where $[S]$ is the vector $(1,n)$ of the sensitivity coefficients dv/dq for combining the POC to the network nodes:

$$
\Delta v_b = [S]\cdot[\Delta q] = \sum_{i=1}^{n} s_i * \Delta q_i
\tag{7}
$$

Depending on the relative coefficients dv/dq, the reactive power variation achieved at different grid nodes thus produces at the POC the voltage variation as in Equation (7). The matrix $[A]$ (n,n) and the vector $[S]$ $(1,n)$ can also be determined by a numerical sensitivity analysis. Indeed, once all the network parameters are established and the reactive power of each DG plant is increased, the consequent voltage variation can be calculated as already discussed in [58]. On the other hand, by starting from the inverse of the electric coupling matrix, the dynamic decoupling matrix is found as in Equation (8). Such a matrix is then composed by the coefficients dv/dq; thus its definition is then given by Equation (9):

$$[DD] = [A]^{-1} \tag{8}$$

$$[\Delta v] = [DD] \cdot [\Delta q] \tag{9}$$

To finally calculate the constants values for the two PI controls, a traditional synthesis is sufficient, once the cascade system is determined. By observing Figure 1, the latter is constituted by the capability matrix, the decoupling matrix, the reactive power regulators, and finally the AVRs or SFCs.

3.2. Traditional Power Plants

The grid topology for a traditional power plant based on fossil fuel is offered in Figure 4, while Figure 5 shows the equivalent electrical model. By comparing the two topologies (i.e., Figure 3 RES versus Figure 4 traditional), the main difference is made manifest: the MV distribution grid in the RES power plant case. In the traditional power plant case, each generator is connected to the main busbar by the only generator transformer, which is characterized by a reactance x_{ti} [49]. This assumption is true even in the case of a generator directly connected to the main busbar, where the compound action provides x_{ti}, i.e., the equivalent reactance introduced by the control. Therefore, it is possible to categorize the traditional power plant as a subcase of the RES power plant case, where the internal distribution network is merely given by the reactance of generator transformers x_{ti}. In such a way, the electrical coupling of generators is only determined by the reactance of generator transformers x_{ti} and the equivalent reactance of upstream network x_{cc}, as clarified in [49]. In this perspective, the network topology represented in Figure 3 and its mathematical model constitutes the general case, albeit it is introduced for the RES case. As a matter of fact, this representation can describe a production power plant either based on RES or on traditional fossil sources. Therefore, the application of control strategy on traditional power plant is also a particular case of the RES production power plant.

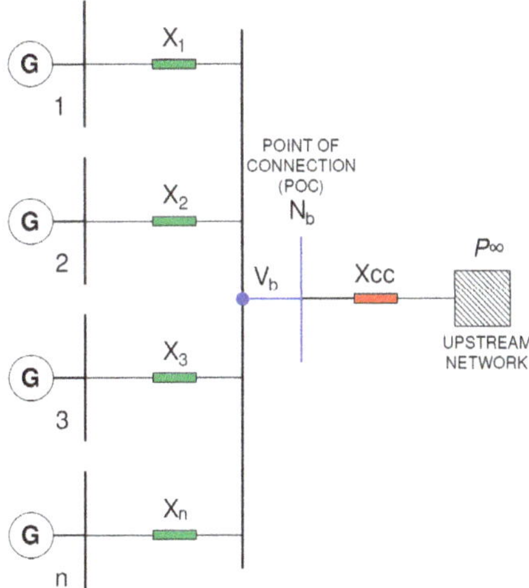

Figure 4. The typical topology of a traditional power plant: several generators parallel-connected to a busbar.

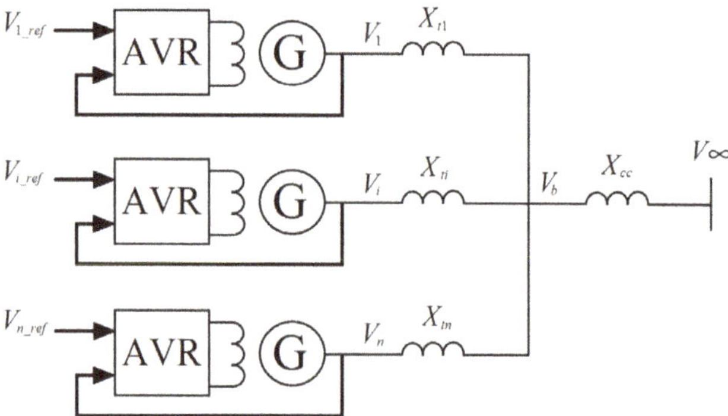

Figure 5. Equivalent electrical model.

For the traditional power plant case, the model can be attained by considering Figure 5. In such a case, the relation between generator voltage variations and reactive power variations is expressed with the algebraic relations (10):

$$\begin{cases} \Delta v_1 = x_{t1} \cdot \Delta q_1 + x_{cc} \cdot \sum_{i=1}^{n} \Delta q_i \\ \vdots \\ \Delta v_i = x_{ti} \cdot \Delta q_i + x_{cc} \cdot \sum_{i=1}^{n} \Delta q_i \\ \vdots \\ \Delta v_n = x_{tn} \cdot \Delta q_n + x_{cc} \cdot \sum_{i=1}^{n} \Delta q_i \end{cases} \tag{10}$$

where the quantities are expressed in the per unit notation. Particularly, x_{ti} is the reactance of the i-th generator transformer, while x_{cc} is the equivalent reactance of the upstream network. The symbol v_i represents the voltage at the terminals of i-th generator and q_i the reactive power of the i-th generator. Finally, v_b is the voltage at POC to the transmission network. By expressing Equation (10) in matrix form, the important Equation (11) is determined. For a traditional electric plant, an additional important result is provided in Equation (12), where the elements of the dynamic decoupling matrix are clearly defined by the reactance x_{ti} and x_{cc} [49].

$$\begin{vmatrix} \Delta v_1 \\ \vdots \\ \Delta v_i \\ \vdots \\ \Delta v_n \end{vmatrix} = \begin{bmatrix} x_{t1} + x_{cc} & x_{cc} & x_{cc} \\ & \vdots & \\ x_{cc} & x_{ti} + x_{cc} & x_{cc} \\ & \vdots & \\ x_{cc} & x_{cc} & x_{tn} + x_{cc} \end{bmatrix} \begin{vmatrix} \Delta q_1 \\ \vdots \\ \Delta q_i \\ \vdots \\ \Delta q_n \end{vmatrix} \tag{11}$$

$$DD_{i,j} = \begin{cases} x_{cc} & \text{if} \quad i \neq j \\ x_{ti} + x_{cc} & \text{if} \quad i = j \end{cases} \tag{12}$$

4. Case Studies

Currently, the proposed algorithm results are already implemented in Italian transmission systems (involving coal-fired and combined cycle gas power stations rated above 100 MVA) to achieve a coordinated production of reactive power [48]. In this section, the experimental data collected from the field for some traditional power plants equipped with a SART apparatus and the simulations for a large PV power plant are reported and compared. Three different power plant configurations are considered. In each of them, tests have been conducted applying at time $t = 35$ s a step in the q_{liv} signal. Simulations have been carried out adopting the proposed control system, according to the mathematical model presented in the third chapter. The mathematical model of the control has been implemented in DOME (a Python based simulation tool) [59], together with the models of the networks for the three cases. Three additional files (Case A, Case B and Case C) are made available as supplementary materials. Datasets and experimental details are in these files.

4.1. Case A) 940 MVA Combined-Cycle Power Plant

The combined-cycle power plant is composed of two identical groups, where each one has a synchro generator-gas turbine (300 MVA) and a synchro generator-steam turbine (170 MVA).

The considered power plant is connected to the Italian 400 kV transmission grid, whilst its topology is depicted in Figure 6. The effectiveness of the control (essentially a PI) is checked by analyzing the step response on the controlled variable reference (the reactive power level, q_{liv}). The obtained results are reported in Figures 7–12. Particularly, the correspondence between the simulated trends and the trends measured in the real plants can validate the proposed mathematical model. Such a correspondence between simulation and experimental data accredits the suitability of the adopted mathematical model.

By observing the last figures, it is possible to see the different level of reactive power reached by the groups two and four; this aspect obviously depends on the different size of groups two and four (170 MVA) in respect of the two twin gas generators of 300 MVA. Particularly, the SART system reactive requests are proportional to the reactive power capability of each generator. Each group changes its generated reactive power according to its capability curve and its generated active power (not shown). In the simulation, a little difference has been considered in the capability of groups one and three, while all active powers are kept constant during the simulations.

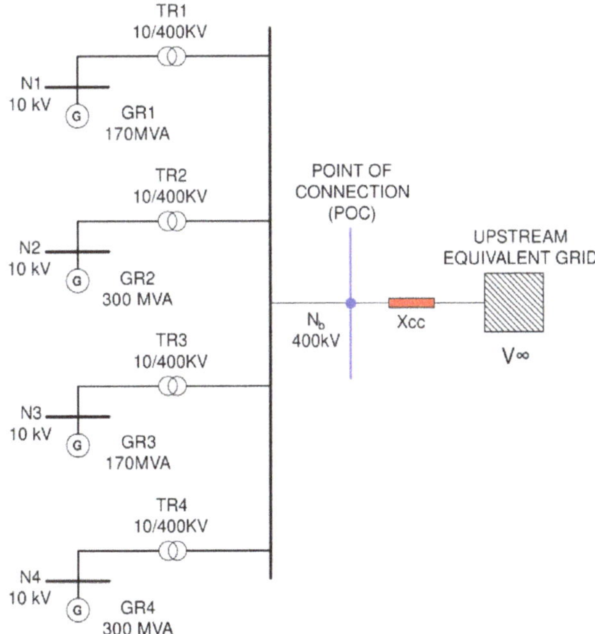

Figure 6. The topology of plant A.

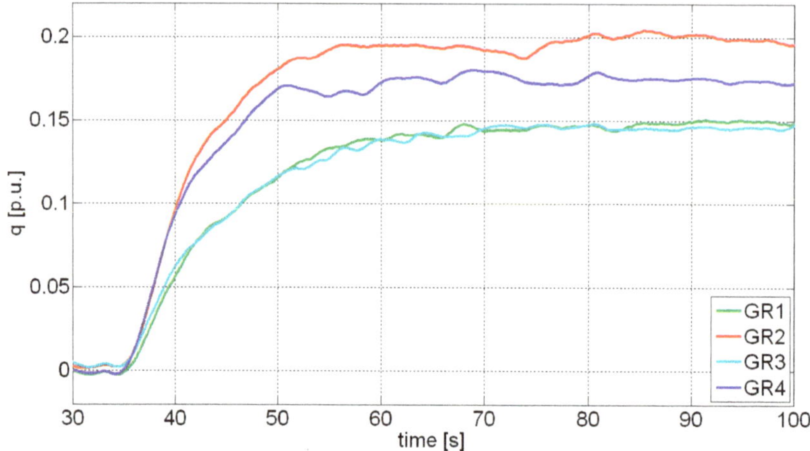

Figure 7. Generators reactive powers response (in p.u.) to a step in the reactive power level reference (experimental data).

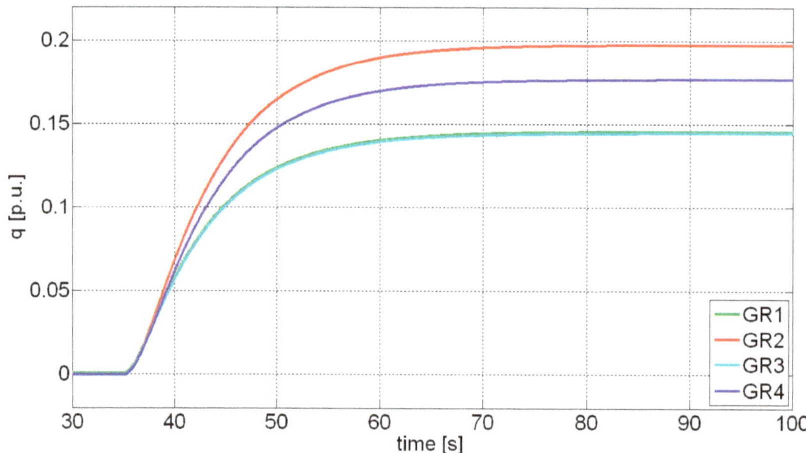

Figure 8. Generators reactive powers response (in p.u.) to a step in the reactive power level reference (simulated data).

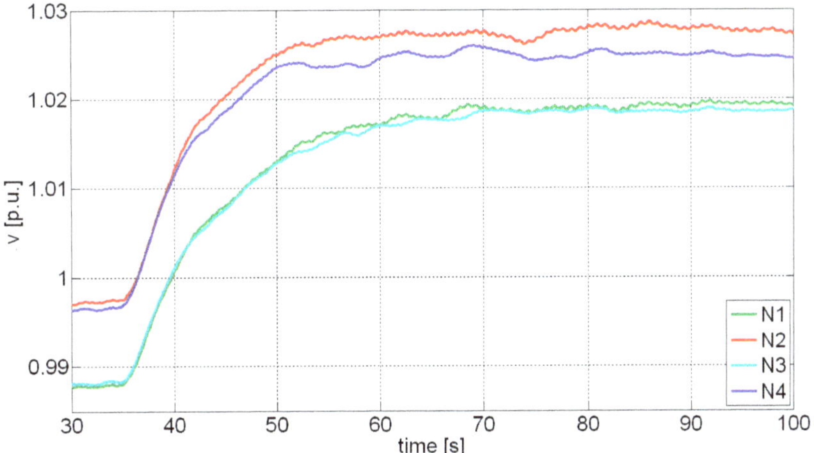

Figure 9. Voltage profile (in p.u.) at the different nodes of generation (experimental data).

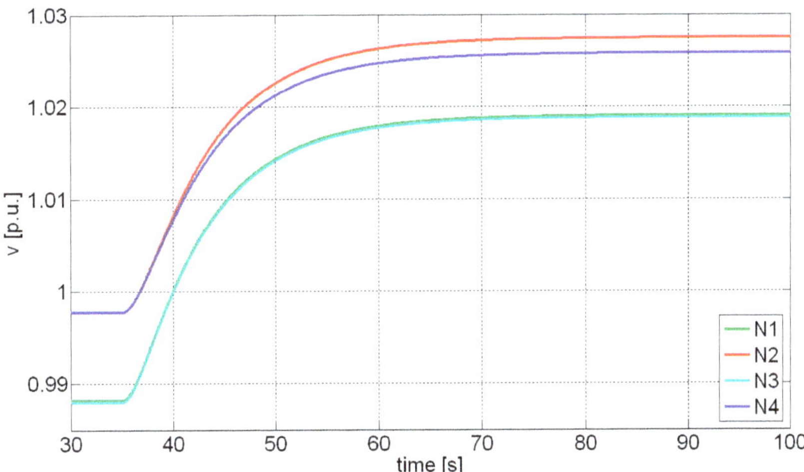

Figure 10. Voltage profile (in p.u.) at the different nodes of generation (simulated data).

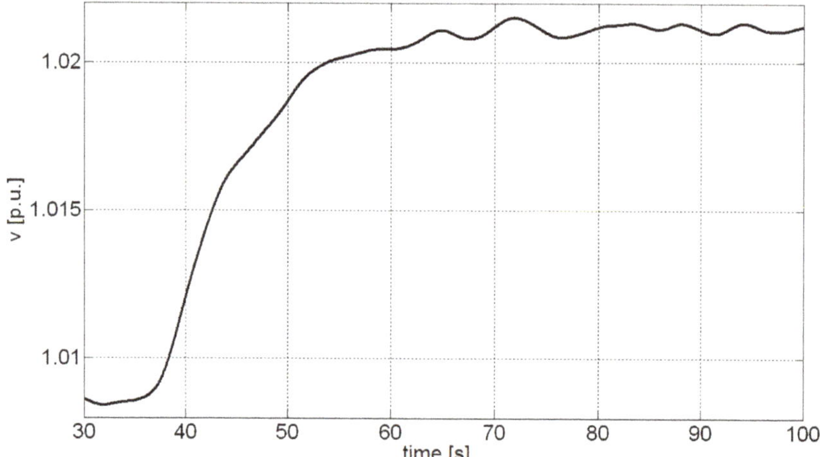

Figure 11. Voltage profile (in p.u.) at the POC (experimental data).

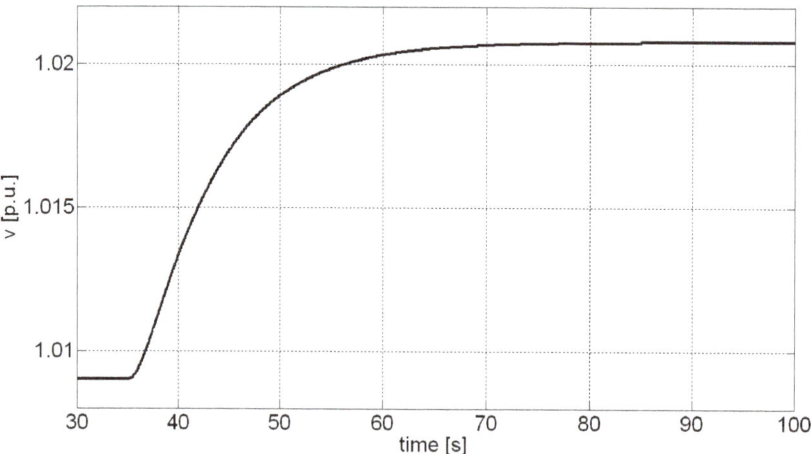

Figure 12. Voltage profile (in p.u.) at the POC (simulated data).

4.2. Case B) Synchronous Condenser of 160 MVA

This case considers a synchronous condenser rated 160 MVA, whose aim is mainly the reactive power production to compensate for voltage perturbations in the transmission network. The plant topology is depicted in Figure 13.

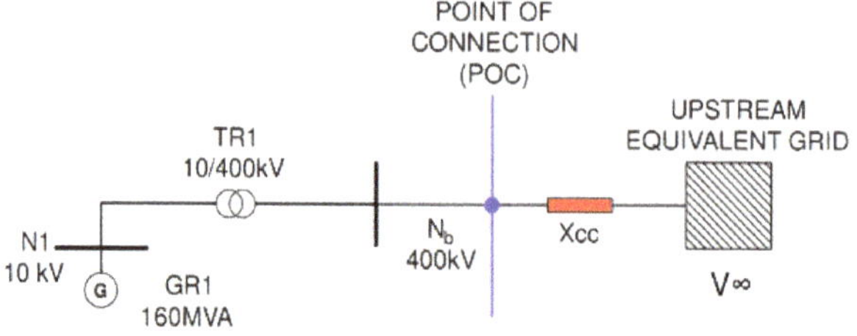

Figure 13. The topology of plant B.

The obtained results are reported in the Figures 14–17. In this case, the mathematical model is very simple. Experimental data and simulations are shown both for voltages and reactive powers. The correspondence between experimental data and simulations is shown as a proof of the correctness of the proposed mathematical model.

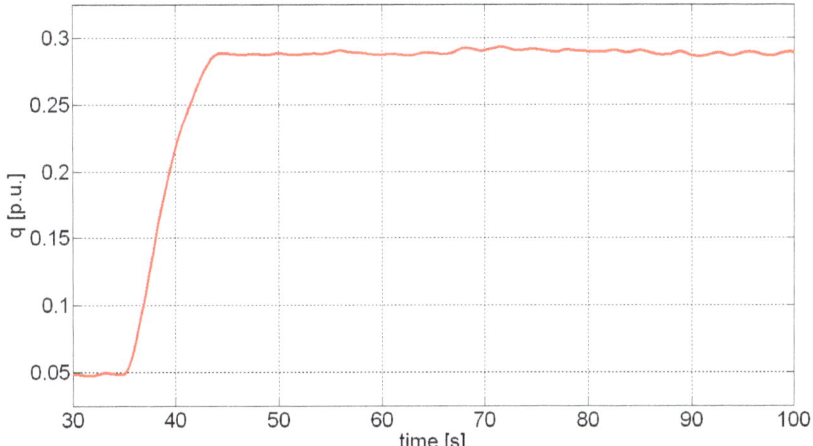

Figure 14. Reactive power profile (in p.u.) of a generator (experimental data).

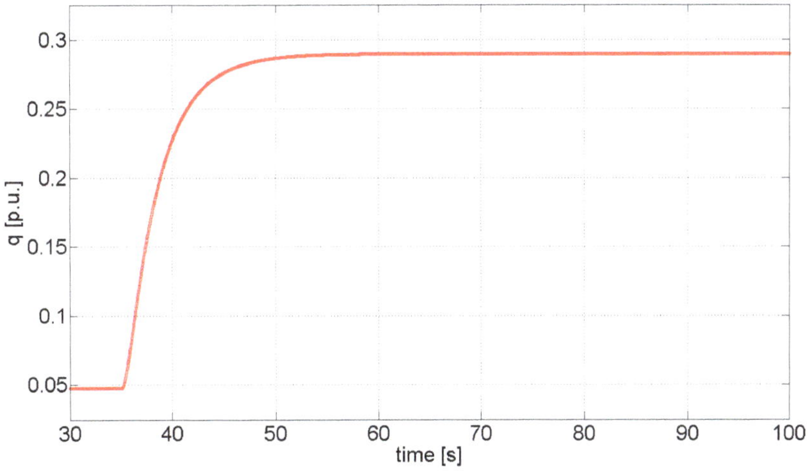

Figure 15. Reactive power profile (in p.u.) of a generator (simulated data).

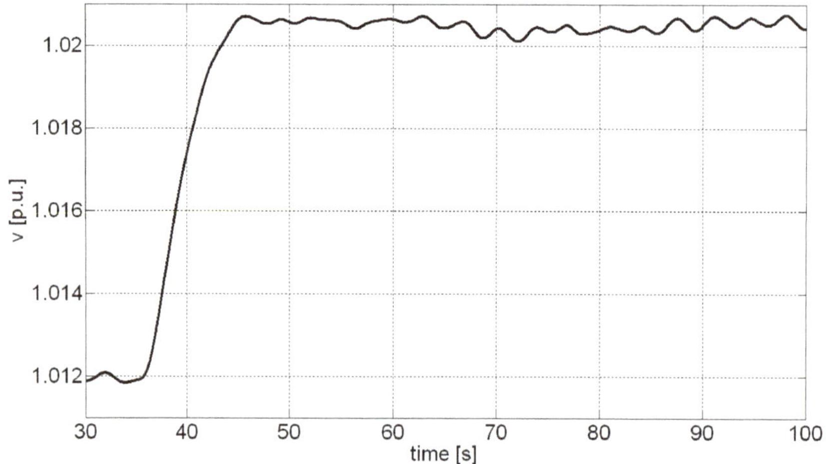

Figure 16. Voltage profile (in p.u.) at the POC (experimental data).

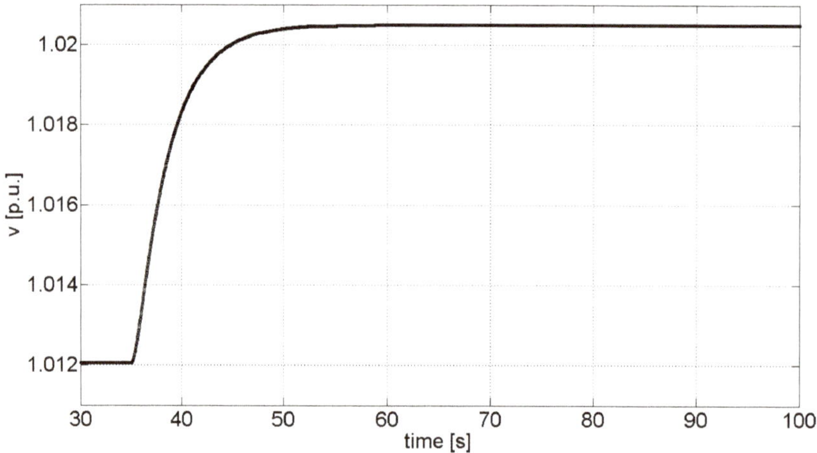

Figure 17. Voltage profile (in p.u.) at the POC (simulated data).

4.3. Case C) PV Power Plant of 46.8 MWp

The considered RES power plant [56] is a large PV plant with a peak power of 46.8 MWp. Particularly, the plant is composed of 82 photovoltaic subfields, each one rated about 0.57 MWp. Every subfield has its own inverter with a rated power of 0.5 MVA, connected to the 20 kV-internal distribution network by a transformer. The topology of the considered network is depicted in Figure 18, whereas all components are interfaced by using two different types of cables (in Figure 18 the two different types are drawn with different style according to Table 1). The lengths of all cables are in the order of hundreds of meters. All lengths have been taken into account in the mathematical model.

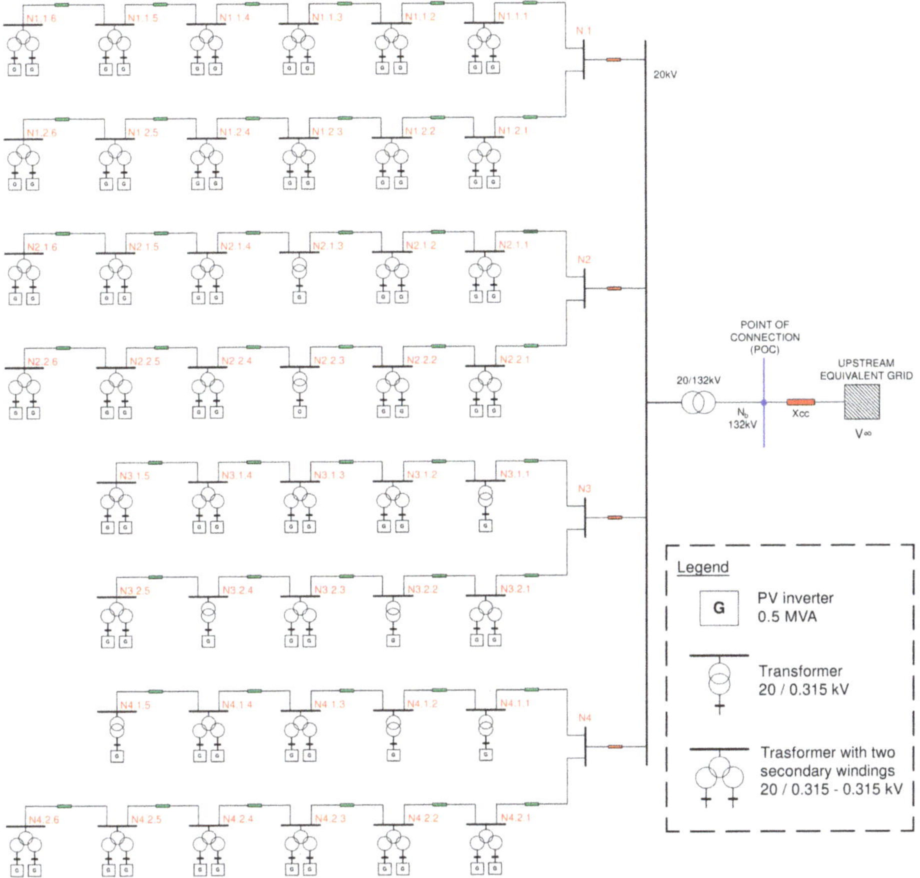

Figure 18. The topology of photovoltaic (PV) power plant, plant C.

Table 1. Cable types.

Line Type	Graphic Symbol	R	X	C
		(Ω/km)	(Ω/km)	(μF/km)
RG7H1R $3 \times 1 \times 240$ mm^2		0.0754	0.12	0.25
RG7H1R $3 \times 1 \times 120$ mm^2		0.153	0.11	0.32

A single 132/20 kV transformer is used to step-up the voltage to the final HV bus, thus the single POC to the HV network. From the transmission grid point of view, the POC is seen as a unique generator similar to a traditional power plant. This PV power plant is traditionally operated at unity power factor, which is no longer considered satisfactory by the TSO. The presence of several generators based on inverters suggests the possibility of using their reactive capability for supporting the HV network voltage control through the proposed control. Additionally, in this PV power plant, a step in the reactive power level is applied as in the previous cases, while only simulation results are shown. In this regard, Figure 19 reports the p.u. reactive power (i.e., rated power as basis) of 6 representative PV inverters, where in total they amount to 82 unities. Each generator injects an amount of reactive power according to its rated power and its residual capability; therefore, the differences are almost negligible. By considering only one size for the inverters, the results are coherent. The voltage outputs of the considered inverters are shown in Figure 20, while the voltage profile at the POC is highlighted in Figure 21. The time responses show stable behaviors, whilst the dynamics result in accordance to the expected time constants. Finally, it possible to highlight the absence of cross dynamics disturbances between the different PV subfields.

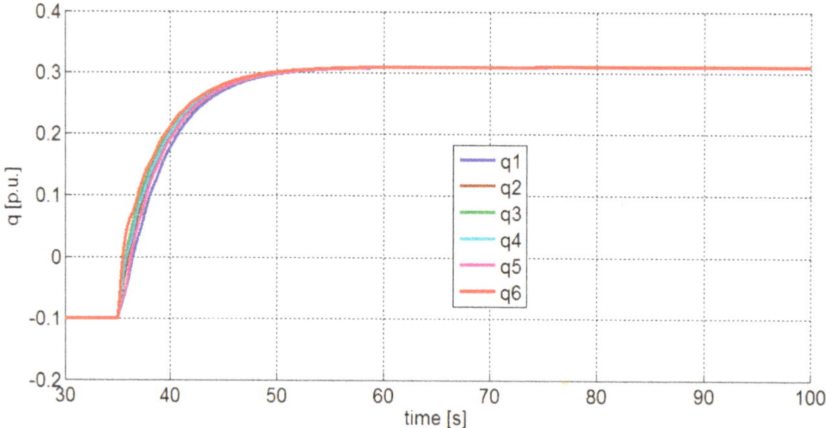

Figure 19. Reactive power profile (in p.u.) of six representative PV inverters (simulated data).

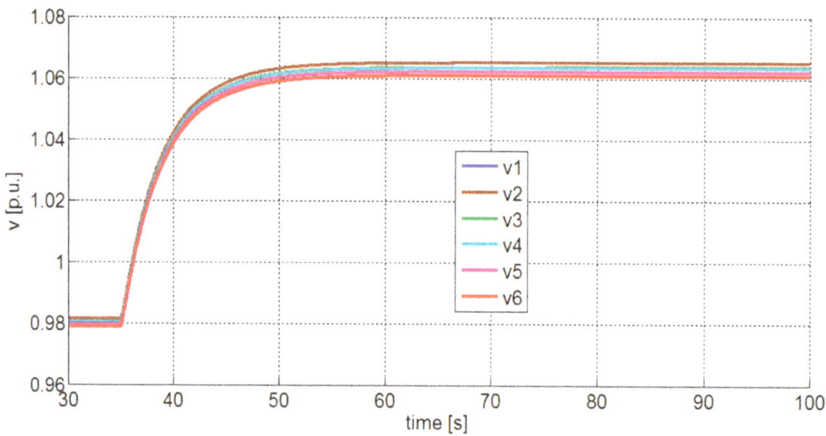

Figure 20. Voltage profile (in p.u.) at the six representative PV inverters output (simulated data).

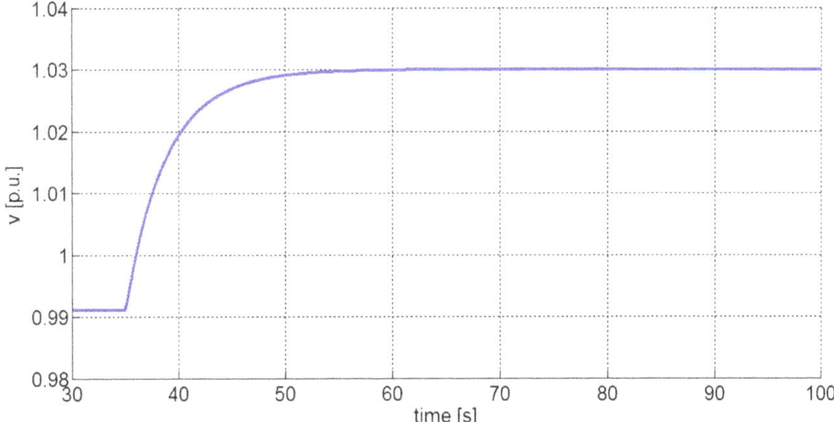

Figure 21. Voltage profile (in p.u.) at the POC (simulated data).

5. Results and Discussion

By focusing on Case A and Case B, the good correspondence between simulated results and experimental data validates the mathematical models, then the way for representing power distribution grid and control architecture in DOME platform. This verified methodology can be then applied to the RES power plant (i.e., Case C), where the results are indeed consequent and thus capable of well-representing the dynamics of this kind of controlled power plant. It is interesting to highlight that both simulations and experimental data show a first-order behavior in presence of a reactive power reference step. On the other hand, the controlled voltage can reach the no-error steady-state condition without oscillations. In such a way, the group of generators (whether in parallel configuration or in star configuration) behaves as a single generator in a primary voltage regulation or as a group when coordinated by means of a secondary voltage regulation. Furthermore, Case C actually shows a behavior similar of what is observable in the presence of a standard power plant operating for a secondary voltage regulation. The latter is an important aspect in which the authors are interested in drawing attention. Finally, the proposed control methodology is not only applicable to the example of this paper but also it can be used when small–medium generators are interconnected to whichever distribution grid's topology.

6. Conclusions

The proposed work has presented a coordinated voltage and reactive power control architecture. The control algorithm has been originated from well-known techniques, already widely used in the transmission network of many national systems for conventional power plants. The novelty is the capability of the algorithm in being adaptable to different technology, size, and topology of generating plants connected to the HV network. A mathematical model of the control system has been discussed and then implemented in DOME (a python-based simulation tool). Firstly, this implementation has been used to simulate the dynamical behavior of traditional power plants, and experimental data have been compared to validate the control system model. Thus, the validated control system model has been applied to a large PV power plant, where numerical simulations have verified the behavior. The performed simulations have demonstrated how a large controlled PV field exhibits an asymptotically stable behavior, in terms of voltage and reactive power. Moreover, the PV dynamics (i.e., time constants) are fully comparable to the ones of large traditional power plants, which are involved in coordinated voltage and reactive power controls in HV networks. Therefore, the reactive capability of PV power plants and large power stations can be synergistically exploited, while keeping a uniform dynamic performance.

The proposed control scheme is finally characterized by several advantages: it is capable of ensuring a fast response, while performing a perfect tracking of both the HV bus voltage and RES reactive powers. In addition, there is no steady-state error and the system dynamics does not highlight oscillations. Each generator participates in the control by sharing its reactive power (in absorption or injection), where the provided quota is proportional to its capability at the point of operation. An important note to be highlighted regards the communication data. The proposed control strategy is based on the communication of a single control signal (i.e., the reactive power level) to all the generators involved, thus making its implementation rather simple. The transmitted q_{liv} value is coherent to the reactive power level's control signal that is already in use in the coordinated voltage and reactive power control of transmission networks. This makes the proposed control architecture fully compatible with existing ones. As a further development, authors are investigating how this control architecture can be scaled and integrated into distribution networks. In a glance, the proposed control is characterized by the two important pros: a) only one signal is requested for regulating multiple-generators and b) it is strictly alike to the secondary control already used in transmission system, thus consequently it is compatible. The proposed solution is feasible when promoting the integration of large PV systems in the HV networks' control strategies (i.e., voltage and reactive power).

Supplementary Materials: The following are available online at http://www.mdpi.com/1996-1073/13/10/2441/s1, File S1: Case A.dm, File S2: Case B.dm, File S3: Case C.dm. Three additional files (Case A, Case B and Case C) are made available during the submission process. Full datasets and full experimental details are provided in these three files. The three files are the inputs for performing simulations then reproducing the results in DOME, the Python based simulation tool used in this paper.

Author Contributions: Conceptualization, M.C., D.B. and G.S.; Data curation, M.C. and R.C.; Formal analysis, R.C.; Methodology, D.B. and G.S.; Software, M.C.; Supervision, G.S.; Validation, D.B.; Visualization, R.C.; Writing—original draft, R.C.; Writing—review & editing, M.C., D.B. and G.S. All authors have read and agreed to the published version of the manuscript.

Funding: This research received no external funding.

Conflicts of Interest: The authors declare no conflict of interest.

Abbreviations

AVR	Automatic Voltage Regulator
BVR	Busbar Voltage Regulator
DD	Dynamic Decoupling
DGs	Distributed Generators
GRPRs	Generator Reactive Power Regulators
HV	High Voltage
LV	Low Voltage
MIMO	Multiple-Input-Multiple-Output
MV	Medium Voltage
PI	Proportional-Integral
POC	Point Of Connection
PV	Photovoltaic
RES	Renewable Energy Sources
RPR	Reactive Power Regulator
SFC	Static Frequency Converter
SISO	Single-Input-Single-Output
STATCOM	Static Synchronous Compensator
SVR	Secondary Voltage Regulation
TSO	Transmission System Operator
VPP	Virtual Power Plants
VSC	Voltage Source Converter
VSN	Virtual Smart Network

References

1. van Wyk, A.A.L.; Khan, M.A.; Barendse, P. Impact of over/under and voltage unbalanced supplies on Energy-Efficient motors. In Proceedings of the 2011 IEEE International Electric Machines & Drives Conference (IEMDC), Niagara Falls, ON, Canada, 15–18 May 2011; pp. 1380–1385.
2. Jahmeerbacus, I.; Bhurtun, C. Energy efficiency and power quality issues of AC voltage controllers in instant water heaters. In Proceedings of the 2012 20th Domestic Use of Energy Conference, Cape Town, South Africa, 3–4 April 2012; pp. 139–144.
3. Lipsky, A.; Braunstein, A.; Miteva, N.; Slonim, M. Electric power quality and efficiency of power supply. In Proceedings of the 21st IEEE Convention of the Electrical and Electronic Engineers in Israel. Proceedings (Cat. No.00EX377), Tel-Aviv, Israel, 11–12 April 2000; pp. 222–225.
4. Fusco, G.; Russo, M. *Adaptive Voltage Control System*; Springer: Berlin/Heidelberg, Germany, 2007.
5. Chiandone, M.; Sulligoi, G.; Massucco, S.; Silvestro, F. Hierarchical voltage regulation of transmission systems with renewable power plants: An overview of the Italian case. In Proceedings of the 3rd Renewable Power Generation Conference (RPG 2014), Naples, Italy, 24–25 September 2014; pp. 1–5.
6. Mousavi, O.A.; Cherkaoui, R. *Literature Survey on Fundamental Issues of Voltage and Reactive Power Control*; École polytechnique fédérale de Lausanne: Lausanne, Switzerland, 2011.
7. Mohandes, B.; el Moursi, M.S.; Hatziargyriou, N.D.; el Khatib, S. A review of power system flexibility with high penetration of renewables. *IEEE Trans. Power Syst.* **2019**, *34*, 3140–3155. [CrossRef]
8. Sun, H.; Guo, Q.; Qi, J.; Ajjarapu, V.; Bravo, R.; Chow, J.; Li, Z.; Moghe, R.; Nasr-Azadani, E.; Tamrakar, U.; et al. Review of challenges and research opportunities for voltage control in smart grids. *IEEE Trans. Power Syst.* **2019**, *34*, 2790–2801. [CrossRef]
9. Vita, V.; Alimardan, T.; Ekonomou, L. The impact of distributed generation in the distribution networks' voltage profile and energy losses. In Proceedings of the 9th IEEE European Modelling Symposium on Mathematical Modelling and Computer Simulation, Madrid, Spain, 6–8 October 2015; pp. 260–265.
10. Ceaki, O.; Seritan, G.; Vatu, R.; Mancasi, M. Analysis of power quality improvement in smart grids. In Proceedings of the 10th International Symposium on Advanced Topics in Electrical Engineering (ATEE), Bucharest, Romania, 23–25 March 2017; pp. 797–801.
11. Nieto, A.; Vita, V.; Maris, T.I. Power quality improvement in power grids with the integration of energy storage systems. *Int. J. Eng. Res. Technol.* **2016**, *5*, 438–443.
12. European Commission. *EU Energy in Figures—Statistical Pocketbook 2016*; Publications Office of the European Union: Brussels, Belgium, 2016.
13. The International Renewable Energy Agency. *Renewable Energy Statistics 2016*; IRENA: Abu Dhabi, UAE, 2016.
14. Renewable Energy Policy Network for the 21st Century (REN21). *Renewables 2014 Global Status Report*; REN21: Paris, France, 2014.
15. GSE S.p.A. *Rapporto Statistico 2014—Solare Fotovoltaico*; GSE S.p.A.: Rome, Italy, 2015.
16. Petinrin, J.O.; Shaaban, M. Impact of renewable generation on voltage control in distribution systems. *Renew. Sustain. Energy Rev.* **2016**, *65*, 770–783. [CrossRef]
17. Pfeiffer, R. Brief status on CENELEC standards related to Connection Network Code. In Proceedings of the 3rd Grid Connection Stakeholder Committee Meeting, Brussels, Belgium, 8 September 2016.
18. Bründlinger, R. Advanced smart inverter and DER functions Requirements in latest European Grid Codes and future trends. In Proceedings of the Solar Canada 2015 Conference, Toronto, ON, Canada, 7–8 December 2015.
19. ENTSO-E. *Implementation Guideline for Network Code "Requirements for Grid Connection Applicable to all Generators"*; ENTSO-E: Brussels, Belgium, 2013.
20. The European Commission. *Commission Regulation (EU) 2016/631 of 14 April 2016 Establishing a Network Code on Requirements for Grid Connection of Generators*; Official Journal of the European Union: Brussels, Belgium, 2016.
21. Arrinda, J.; Barrena, J.A.; Rodríguez, M.A.; Guerrero, A. Analysis of massive integration of renewable power plants under new regulatory frameworks. In Proceedings of the 2014 International Conference on Renewable Energy Research and Application (ICRERA), Milwaukee, WI, USA, 19–22 October 2014; pp. 462–467.
22. Varma, R.K.; Salama, M. Large-scale photovoltaic solar power integration in transmission and distribution networks. In Proceedings of the 2011 IEEE Power and Energy Society General Meeting, Detroit, MI, USA, 24–28 July 2011; pp. 1–4.

23. Delfino, F.; Procopio, R.; Rossi, M.; Ronda, G. Integration of large-size photovoltaic systems into the distribution grids: A P-Q chart approach to assess reactive support capability. *IET Renew. Power Gener.* **2010**, *4*, 329–340. [CrossRef]

24. Tyll, H.K.; Schettle, F. Historical overview on dynamic reactive power compensation solutions from the begin of AC power transmission towards present applications. In Proceedings of the 2009 IEEE/PES Power Systems Conference and Exposition, Seattle, WA, USA, 15–18 March 2009; pp. 1–7.

25. Ghasemi, M.A.; Parniani, M. Prevention of distribution network overvoltage by adaptive droop-based active and reactive power control of PV systems. *Electric Power Syst. Res.* **2016**, *133*, 313–327. [CrossRef]

26. Hashemi, M.; Agelidis, V. Evaluation of voltage regulation mitigation methods due to high penetration of PV generation in residential areas. In Proceedings of the 2013 International Conference on Renewable Energy Research and Applications (ICRERA), Madrid, Spain, 20–23 October 2013; pp. 1180–1189.

27. Hasheminamin, M.; Agelidis, V.G.; Ahmadi, A.; Siano, P.; Teodorescu, R. Single-point reactive power control method on voltage rise mitigation in residential networks with high PV penetration. *Renew. Energy* **2018**, *119*, 504–512. [CrossRef]

28. Rallabandi, V.; Akeyo, O.M.; Ionel, D.M. Modeling of a multi-megawatt grid connected PV system with integrated batteries. In Proceedings of the 2016 IEEE International Conference on Renewable Energy Research and Applications (ICRERA), Birmingham, UK, 20–23 November 2016; pp. 1146–1151.

29. Olival, P.C.; Madureira, A.G.; Matos, M. Advanced voltage control for smart microgrids using distributed energy resources. *Electr. Power Syst. Res.* **2017**, *146*, 132–140. [CrossRef]

30. Jayasekara, N.; Wolfs, P.; Masoum, M.A.S. An optimal management strategy for distributed storages in distribution networks with high penetrations of PV. *Electr. Power Syst. Res.* **2014**, *116*, 147–157. [CrossRef]

31. Ko, H.; Bruey, S.; Jatskevich, J.; Dumont, G.; Moshref, A. A PI Control of DFIG-Based Wind Farm for Voltage Regulation at Remote Location. In Proceedings of the 2007 IEEE Power Engineering Society General Meeting, Tampa, FL, USA, 24–28 June 2007; pp. 1–6.

32. Baalbergen, F.; Gibescu, M.; van der Sluis, L. Voltage stability consequences of decentralized generation and possibilities for intelligent control. In Proceedings of the 2010 IEEE PES Innovative Smart Grid Technologies Conference Europe (ISGT Europe), Gothenburg, Sweden, 11–13 October 2010; pp. 1–8.

33. Zhao, J.; Li, X.; Hao, J.; Lu, J. Reactive power control of wind farm made up with doubly fed induction generators in distribution system. *Electr. Power Syst. Res.* **2010**, *80*, 698–706. [CrossRef]

34. El-Fouly, T.H.M.; El-Saadany, E.F.; Salama, M.M.A. Voltage regulation of wind farms equipped with variable-speed doubly-fed induction generators wind turbines. In Proceedings of the 2007 IEEE Power Engineering Society General Meeting, Tampa, FL, USA, 24–28 June 2007; pp. 1–8.

35. Qin, N.; Bak, C.L.; Abildgaard, H. Automatic voltage control system with market price employing large wind farms. *Electr. Power Syst. Res.* **2018**, *157*, 93–105. [CrossRef]

36. Mastromauro, R.A.; Orlando, N.A.; Ricchiuto, D.; Liserre, M.; Dell'Aquila, A. Hierarchical control of a small wind turbine system for active integration in LV distribution network. In Proceedings of the 2013 International Conference on Clean Electrical Power (ICCEP), Alghero, Italy, 11–13 June 2013; pp. 426–433.

37. Koraki, D.; Strunz, K. Wind and solar power integration in electricity markets and distribution networks through service-centric virtual power plants. *IEEE Trans. Power Syst.* **2018**, *33*, 473–485. [CrossRef]

38. Unger, D.; Spitalny, L.; Myrzik, J.M.A. Voltage control by small hydro power plants integrated into a virtual power plant. In Proceedings of the 2012 IEEE Energytech, Cleveland, OH, USA, 29–31 May 2012; pp. 1–6.

39. Dielmann, K.; van der Velden, A. Virtual power plants (VPP)—A new perspective for energy generation? In Proceedings of the 9th International Scientific and Practical Conference of Students, Post-graduates Modern Techniques and Technologies, 2003, MTT 2003, Tomsk, Russia, 7–11 April 2003.

40. Pudjianto, D.; Ramsay, C.; Strbac, G. Virtual power plant and system integration of distributed energy resources. *IET Renew. Power Gener.* **2007**, *1*, 10–16. [CrossRef]

41. Ricerca Sistema Energetico (RSE). *Definizione dei Concetti e Delle Architetture di Gestione, Controllo e Comunicazione di Microreti e Virtual Power Plant*; Progetto RdS—Trasmissione e Distribuzione; RSE: Milan, Italy, 2006.

42. Arcidiacono, V. Automatic Voltage and Reactive Power Control in Transmission Systems. In Proceedings of the 1983 CIGRE-IFAC Symposium, Florence, Italy, 26–28 September 1983. Survey paper.

43. Paul, J.P.; Leost, J.Y.; Tesseron, J.M. Survey of the Secondary Voltage Control in France: Present Realization and Investigations. *IEEE Trans. Power Syst.* **1987**, *2*, 505–511. [CrossRef]

44. Sancha, J.L.; Fernandez, J.L.; Cortes, A.; Abarca, J.T. Secondary voltage control: Analysis, solutions and simulation results for the Spanish transmission system. *IEEE Trans. Power Syst.* **1996**, *11*, 630–638. [CrossRef]
45. Automatic System for Voltage Regulation (SART) for Electric Power Plant, GRTN Document nr. DRRPX03019. Available online: http://www.terna.it/LinkClick.aspx?fileticket=irZ1FD%2BYxUE%3D&tabid=106&mid=468 (accessed on 11 April 2020).
46. Lagonotte, P.; Sabonnadiere, J.C.; Leost, J.Y.; Paul, J.P. Structural analysis of the electrical system: Application to secondary voltage control in France. *IEEE Trans. Power Syst.* **1989**, *4*, 479–486. [CrossRef]
47. Sancha, J.L.; Fernandez, J.L.; Martinez, F.; Salle, C. Spanish practices in reactive power management and voltage control. In Proceedings of the IEE Colloquium on International Practices in Reactive Power Control, London, UK, 7 April 1993; pp. 3/1–3/4.
48. Sulligoi, G.; Chiandone, M.; Arcidiacono, V. NewSART automatic voltage and reactive power regulator for secondary voltage regulation: Design and application. In Proceedings of the 2011 IEEE Power and Energy Society General Meeting, San Diego, CA, USA, 24–28 July 2011; pp. 1–7.
49. Arcidiacono, V.; Menis, R.; Sulligoi, G. Improving power quality in all electric ships using a voltage and VAR integrated regulator. In Proceedings of the 2007 IEEE Electric Ship Technologies Symposium, Arlington, VA, USA, 21–23 May 2007; pp. 322–327.
50. TERNA. Sistema Automatico per la Regolazione di Tensione (SART) per Centrali Elettriche di Produzione. Annex A16 to the National Grid Code, GRTN Document nr. DRRPX03019. Available online: http://collaudo.download.terna.it/terna/0000/0105/33.pdf (accessed on 11 April 2020).
51. Tamimi, B.; Cañizares, C.; Bhattacharya, K. Modeling and performance analysis of large solar photo-voltaic generation on voltage stability and inter-area oscillations. In Proceedings of the 2011 IEEE Power and Energy Society General Meeting, San Diego, CA, USA, 24–28 July 2011; pp. 1–6.
52. Vasquez, J.C.; Mastromauro, R.A.; Liserre, J.M. Voltage support provided by a droop-controlled multifunctional inverter. *IEEE Trans. Ind. Electron.* **2009**, *56*, 4510–4519. [CrossRef]
53. Cagnano, A.; De Tuglie, E.; Liserre, M.; Mastromauro, R.A. Online Optimal Reactive Power Control Strategy of PV Inverters. *IEEE Trans. Ind. Electron.* **2011**, *58*, 4549–4558. [CrossRef]
54. Campaner, R.; Chiandone, M.; Arcidiacono, V.; Sulligoi, G.; Milano, F. Automatic voltage control of a cluster of hydro power plants to operate as a virtual power plant. In Proceedings of the 2015 IEEE 15th International Conference on Environment and Electrical Engineering (EEEIC), Rome, Italy, 10–13 June 2015; pp. 2153–2158.
55. Chiandone, M.; Campaner, R.; Arcidiacono, V.; Sulligoi, G.; Milano, F. Automatic voltage and reactive power regulator for wind farms participating to TSO voltage regulation. In Proceedings of the 2015 IEEE Eindhoven PowerTech, Eindhoven, The Netherlands, 29 June–2 July 2015; pp. 1–5.
56. Chiandone, M.; Campaner, R.; Pavan, A.M.; Arcidiacono, V.; Milano, F.; Sulligoi, G. Coordinated voltage control of multi-converter power plants operating in transmission systems. The case of photovoltaics. In Proceedings of the 2015 International Conference on Clean Electrical Power (ICCEP), Taormina, Italy, 16–18 June 2015; pp. 506–510.
57. Gao, B.; Morison, G.K.; Kundur, P. Voltage Stability evaluation using modal analysis. *IEEE Trans. Power Syst.* **1992**, *7*, 1529–1542. [CrossRef]
58. Carvalho, P.M.; Correia, P.F.; Ferreira, L.A. Distributed reactive power generation control for voltage rise mitigation in distribution networks. *IEEE Trans. Power Syst.* **2008**, *23*, 766–772. [CrossRef]
59. Milano, F. A python-based software tool for power system analysis. In Proceedings of the IEEE PES General Meeting, Vancouver, BC, Canada, 21–25 July 2013.

Article

Dynamic Reconfiguration Systems for PV Plant: Technical and Economic Analysis

Giuseppe Schettino, Filippo Pellitteri, Guido Ala, Rosario Miceli, Pietro Romano and Fabio Viola *

Department of Engineering, University of Palermo, 90128 Palermo, Italy; giuseppe.schettino@unipa.it (G.S.); filippo.pellitteri@unipa.it (F.P.); guido.ala@unipa.it (G.A.); rosario.miceli@unipa.it (R.M.); pietro.romano@unipa.it (P.R.)
* Correspondence: fabio.viola@unipa.it

Received: 16 March 2020; Accepted: 15 April 2020; Published: 17 April 2020

Abstract: Solar plants suffer of partial shading and mismatch problems. Without considering the generation of hot spots and the resulting security issues, a monitoring system for the health of a PV plant should be useful to drive a dynamic reconfiguration system (DRS) to solve bottlenecks due to different panels' shading. Over the years different DRS architectures have been proposed, but no suggestions about costs and benefits have been provided. Starting from technical subjects such as differences of the topologies driving the hardware complexity and number of components, this paper identifies the cost of DRS and its lifetime, and based on these issues it provides an economic analysis for a 6 kWp PV plant in different European Union countries, in which the dissimilar incentive policies have been considered.

Keywords: energy policies related to PV power plants; economic analyses; monitoring and case studies

---

1. Introduction

The trend of the last decade to reduce greenhouse gas emissions and decarbonize the energy sources is mainly due to the increase of the temperature of the planet, has driven the development and improvement of technologies for renewable energy sources. As well known, solar energy is the most available renewable energy source in the world with respect to the other sources and an interesting topic of research is the evaluation of its economic convenience. For this reason, in many studies reported in literature, the economic convenience of grid-connected PV systems has been evaluated with respect to other technologies. The traditional approach is based on the use of the "levelized cost" method that represents the per unit value of total costs (i.e., capital, operation and maintenance, fuel) over the economic life of the power plant [1,2]. The notion of economic convenience of PV plants emerged from these studies.

In the European Union (EU), the number of the PV plants is increasing significantly thanks to the policies of the different countries that have led to economic benefits for private citizens and in particular for small size PV plants called "residential plants" (from 1 kWp to 10 kWp) through diverse incentive systems. The international policies to encourage the installation of the PV plants usually consist of an incentive for the total "green energy" produced and in an incentive for the energy injected into the grid only for the grid-connected plants.

As well known, a typical issue of PV plants is the power loss due to the differences of irradiation (partial shading or wrong design) among the cells of the same module or among different modules of an array. This phenomenon, known as mismatch, can generate a considerable power loss of the total system with a consequent economic loss. In [3] an interesting estimation of PV mismatch losses caused by moving clouds is reported. Moreover, this phenomenon can be caused from a fixed obstacle that may have appeared years after the installation of the PV plant. The presence of a fixed obstacle after the installation is a frequent case in residential PV plants.

In order to limit the mismatch phenomenon, monitoring systems are extensively used in renewable energy applications to track the performance of the generation plant. A monitoring system for PV arrays is usually needed to collect power production and performance data as well as weather condition information and relate them [4–6]. These systems allow detecting fault conditions, but they are not effective for estimating the power reduction of a PV plant.

The issue of the different irradiation levels among the cells of a module has been studied in [7], where an investigation on partially shaded modules with different PV cell connections was reported. The authors compared five different connection configurations in order to find the best solution to increase the maximum power production and the fill factor of a module. The same problem has been studied in [8] and [9].

The different irradiance causes also problems in maximum power point tracking (MPPT) algorithms because the P-V curves of the PV module exhibit multiple maximum power points due to the bypass diodes, which are used to exclude the module of an array. In [10], the authors classified MPPT techniques for different PV array configurations. Obviously, each method presents advantages and drawbacks. Again, an interesting MPPT strategy for PV arrays under uniform and non-uniform irradiance condition is described in [11].

A recent solution proposed in the literature to reduce the power losses is the use of a dynamic reconfiguration system (DRS). The DRS allows one to change the configuration of the PV plant in order to increase the power production. Different solutions have been recently proposed in literature to optimize the power output adopting dynamic reconfiguration systems for PV module interconnection [12–19].

An interesting topic about DRS concerns the economic benefits introduced by the use of these systems. In [20] a technical-economical evaluation on the use of a DRS in some EU countries for PV plants is reported. In particular, by considering the incentive policies and others technical aspects of a 3 kW PV plant the NPV have been evaluated for each country taken into account. Nevertheless, in this study the technical aspects and different configurations of the DRS were not considered. For this reason, it is necessary to extend the economic analysis by introducing the real technical considerations of the DRS for different configurations reported in literature.

The aim of this paper is to evaluate the economic benefits accrued by using a DRS in a residential PV plant. First, an economic analysis of different DRSs according to the costs of the components and to the adopted topological schemes, is carried out; to the authors' knowledge, this issue has never been addressed in the technical literature. Architectures involve switching matrix, sensing network and driving circuit, the choice of switches affects the electro-technical and, electrical endurance. As will be shown in the following sections, the choice of a more flexible DRS comprises higher initial cost due to the number of switches required by the adopted architecture, but at the same time a less exploitation and therefore a longer useful life.

In particular, the study takes into account different technical and economic aspects of a PV plant in order to present a complete economic analysis. In other words, the study is focused on the use of different DRS configurations reported in literature, in some EU countries in order to evaluate the performances of the investment. The economic tools are the net present value (NPV) and payback time.

This paper is organized as follows: Section 2 provides a brief description of the DRS topology taken into account in this work and technical considerations to estimate the costs and lifetime of DRS. Section 3 describes the experimental set-up to perform the evaluation of performance of DRSs. In Section 4 the economic data are reported and in Section 5 the economic results are presented. Finally, Section 6 concludes the paper.

2. Dynamic Reconfiguration Systems

A DRS allows changing the connections among PV modules in order to increase the total power production from a PV plant under poor irradiance conditions or other situations, that determine the degradation of the performance. In this way, the hardware complexity of the DRS depends on the possible connections among the modules. Generally, in a reconfiguration algorithm each panel is a considered a node of the dynamic array; the number of the nodes is identified as *m* while *n* switches perform the dynamic connections among the panels. A plant with a high number of panels requires a DRS with a high number of switches in order to connect all the nodes. Thus, a topology with more switches guarantees a high number of possible configurations for connection of the panels. In this section, a brief state of art of the dynamic reconfiguration systems (DRS) and a technical-economic analysis of the four cases studied, are reported.

2.1. Case Studies

In order to carried out a complete technical-economic analysis, four DRS cases have been investigated. The same cases have been studied in [21], but without considering the implications in terms of cost and lifetime of the DRS. The first case under test has been presented in [22]. The authors proposed an optimised switch set (SWS) topology for reconfiguration of PV panels based on a particle swarm optimization (PSO) algorithm. Figure 1 shows the optimised topology structure suggested by the authors, in which there are four lines and ten switches.

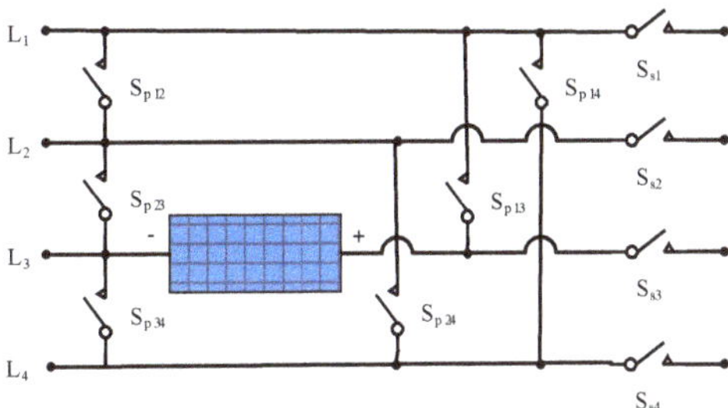

Figure 1. Optimised topology structure proposed in [22], IET: 2018.

From Figure 1, it is possible to observe that the number of switches *n* for each node *m*, is equal to:

$$n_{case1} = m \cdot 10 \tag{1}$$

An interesting low-cost method has been presented in [23]. This method does not require any additional MPPT controllers or sensors and it is based on the use of fuzzy logic (FL) which is used to identify shaded, dirty or faulty panels, to estimate the percentage of shading or dust and to evaluate the minimum and maximum voltage values at which PV panels should be connected/disconnected. The validity of this system has been demonstrated through experimental tests. Figure 2 shows the connection of the system with four panels described in case 2.

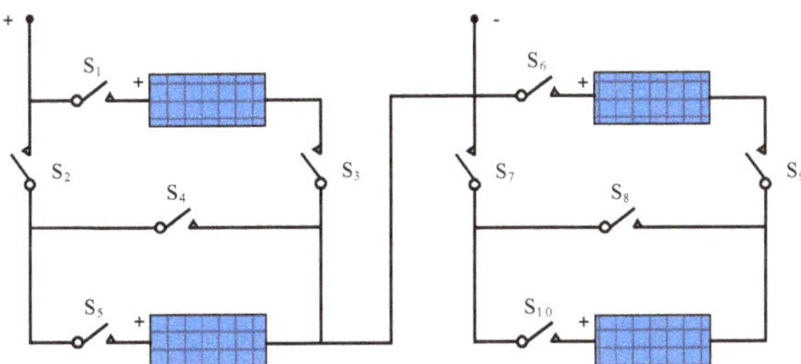

Figure 2. Topology structures obtained with the method proposed in [23], Elsevier: 2015.

The number of switches required for the case 2 can be evaluated as:

$$n_{case2} = m + 6 \tag{2}$$

In [24] a photovoltaic array switching algorithm is presented. This algorithm, in order to find the best configuration of a PV array, is based on the use of only two parameters: the array load voltage and the PV module's temperature. The study has been focused on the evaluation of the performance of four PV modules, as shown in Figure 3.

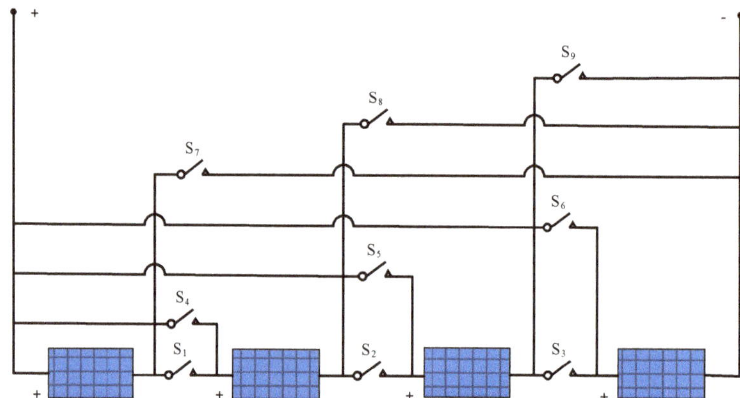

Figure 3. Topology structures obtained with the algorithm proposed in [24], reproduced from the proceedings of the TENCON 2009, IEEE: 2009.

The number of switches necessary in case 3 is equal to:

$$n_{case3} = 3 \times (m - 1) \tag{3}$$

The last case taken into account in this work is presented in [25]. Case 4 is a system configuration approach using an adaptive architecture based on a switching matrix. The adaptive strategy is based on the fact that the switching matrix allows one to rearrange the active PV modules in series into multiple strings to meet the required voltage level. Figure 4 shows the proposed switching matrix of case 4.

Also, in this case, the number of switches of the matrix depends by the number of modules in the PV array. The number of switches can be expressed as:

$$n_{case4} = 4 \times (m) + 2 \times (dc) + 2 \times (inv) \tag{4}$$

where the terms *2dc* and *2inv* represent the switches to connect the PV array with the inverter and the direct current converter. In the next section, the cost estimation analysis is reported.

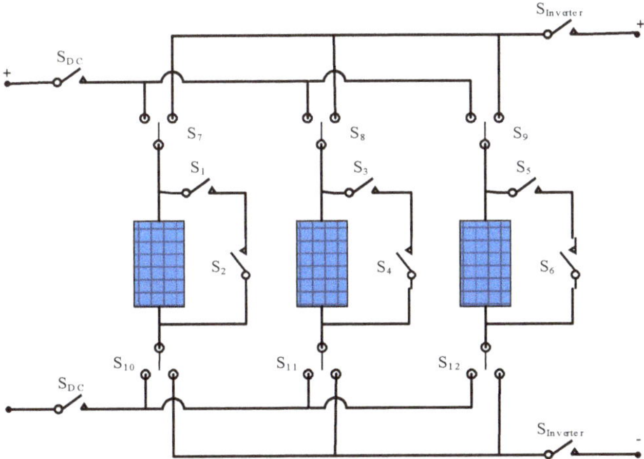

Figure 4. Switching matrix proposed in [25], reproduced from the proceedings of the IECON 2010, IEEE: 2010.

2.2. Costs Estimation of DRS

The cost of each reconfigurator system has been evaluated on the basis of the direct proportionality between the number of switches composing the system and the cost of the technology needed to produce it.

For each of the four considered reconfiguration cases, a cost estimation of the reconfiguration system has been carried out according to the following procedure: for each case, the required amount of components has been evaluated; after that, for each type of electrical component, a specific item which is available on the market has been chosen, compliant with the technical requirements of the system; finally, for each of the selected components, a price is given, as provided by a major distributor of electronics [26].

Generally, the hardware of a dynamic reconfigurator basically consists of three different parts: the switching matrix, the sensing network and the driving circuit. The switching matrix includes all the switches that are used in the reconfiguration system. Taking into account the solutions available in the market, each switch is generally assembled through two parallel-connected devices: an electromechanical relay and a semiconductor device, e.g., a MOSFET [27,28]. A state-solid relay is a valid alternative as well, but it turns out to be more expensive. Whenever the switch has to be put in the off state, the semiconductor switch is closed as first, so that the electromechanical switch can be switched off at a low voltage. When the switch is on, the electromechanical relay guarantees the conduction losses minimization. In this way, the current breaking capability of the electromechanical switch is fully employed, since it is better to be opened at a quite low voltage. The number of parallel-connected MOSFETs arises from the type of electromechanical switches: one MOSFET has to be connected to one single pole single throw (SPST), whereas two MOSFETs have to be connected to one single pole double throw (SPDT). In Table 1 the chosen electromechanical relays are reported along with their respective prices [26,29,30].

Table 1. Electromechanical relays types.

Component	Switch	
	SPDT	SPST
Type	SPDT	SPST
Brand	Finder	Hongfa Europe GMbH
Price (€)	5.22	2.6

In Table 2 the selected MOSFETs and drivers and respective prices are reported as well [26,31,32].

Table 2. MOSFET and driver taken into account.

Component	MOSFET	Driver
Type	n-ch.	Isolated
Brand	IPB08CN10N	MAX845
Price (€)	1.273	3.2

As far as the sensing network, three types of measurements are generally needed: voltage, current and temperature. In all four cases provided in this paper no irradiance sensor is required. The electronics involved in the voltage sensing circuit are normally very cheap, therefore voltage sensors are not considered in the hardware balance for the sake of simplicity. The selected sensors of temperature and current and their price are reported in Table 3. Note that for the current measurement, the selected sensors are compliant with a 6A rated current [26,33,34].

Table 3. Selected sensors for current and temperature.

Component	Sensors	
Type	Temperature	Current
Brand	LM35	LEM
Price (€)	1.393	12.11

Table 4 provides the details concerning the number of the different components which should be used in the four considered reconfiguration cases: PV modules (Np), SPST and SPDT switches (N_{SPST} and N_{SPDT}), MOSFETs, drivers and sensors. The total price has been calculated according to the cost tables previously reported.

Table 4. Components considered for each case.

Cost of Components	Case 1	Case 2	Case 3	Case 4
Np	20	20	20	20
N_{SPST}	200	26	57	44
N_{SPDT}	0	0	0	40
MOSFET	200	26	57	124
Drivers	200	26	57	124
Current sensors	20	0	0	20
Temperature sensors	0	0	20	20
Total price (€)	1 657	184	431	1 148

Note: Each driver is supposed to drive one MOSFET.

2.3. Lifetime Estimation of DRS

The lifetime of each reconfiguration solution has a significant contribution to the overall economical impact, even though in the scientific literature this aspect is generally neglected [35] Regarding this, the most important issue to be addressed is the lifetime of the relays, due to their mechanical characteristics. Both the electrical and the mechanical endurance are reported in the technical datasheets. Indeed, both the electrical and the mechanical behaviour of the relay are affected by the switching operations. More in detail, the electrical endurance, given by the maximum number of cycles recommended to not affect the electrical behaviour of the relay, is usually much shorter than the maximum number of cycles recommended not to affect the mechanical behaviour. Therefore, being first in the time course, only the electrical endurance has been considered in the overall lifetime estimation.

As reported in the technical datasheets, the maximum number of cycles for the selected SPST switches is 10^5, whereas for the selected SPDT switches this value is 60×10^3 [29,30]. In order to evaluate the actual number of switching operations for each reconfiguration case, the specific algorithm as well as the irradiance conditions should be exactly known. Nevertheless, being these data due to the designer in the first case and unpredictable in the second, a simple approach has been here adopted. Considering N_{SPST} and N_{SPDT} the number of SPST and of SPDT respectively, the probability of a switching operation for each of them has been considered to be $1/N_{SPST}$ and $1/N_{SPDT}$. These are meant to be the probability values whenever the algorithm and the irradiance condition lead to a reconfiguration operation. As far as the hours of sunlight and the frequency of reconfiguration are concerned, two "worst case" values have been considered: 16 h of sunlight and 1 reconfiguration every minute. Even though these values are generally peak values across the whole day, these are meant to be the average values, so that a "worst case" situation is taken into account. In Table 5, the number of considered sunlight hours, the number of considered reconfiguration operations per minute and the electrical endurance of SPST and SPDT are given, referenced as E_{SPST} and E_{SPDT}, respectively.

Table 5. Main characteristics of the proposed study case of reconfiguration.

Sunlight hours	16
Reconfigurations Per Minute	1
E_{SPST}	10^5
E_{SPDT}	60×10^3

According to the 16 light hours and one reconfiguration per minute, 350,400 operations are calculated per year, so that the corresponding number of reconfiguration per switch is calculated, according to (5):

$$R_{yrsw,SPST} = R_{yr} \cdot (1/N_{SPST}) \quad R_{yrsw,SPDT} = R_{yr} \cdot (1/N_{SPDT}) \quad N_{yr,SPST} = E_{SPST}/R_{yrsw,SPST} \quad N_{yr,SPDT} = E_{SPDT}/R_{yrsw,SPDT} \tag{5}$$

where: R_{yr} is the number of reconfigurations per year; $R_{yrsw,SPST}$ and $R_{yrsw,SPDT}$ are the number of reconfigurations per year per switch for SPST and SPDT respectively; $N_{yr,SPST}$ and $N_{yr,SPDT}$ express, in terms of number of years, the endurance of SPTS and SPDT switches respectively.

Table 6 reports the data referring to both types of switches and to the four considered cases of reconfiguration, obtained from (5). Note that the number of total reconfigurations R_{yr} has been considered the same for all the cases.

Table 6. Estimated number of reconfigurations per year and endurability of the switches in the four cases, for the "worst case" condition.

Number of Reconfigurations	Case 1	Case 2	Case 3	Case 4
$1/N_{SPST}$	0.005	0.038	0.018	0.023
$1/N_{SPDT}$				0.025
R_{yr}	350,400	350,400	350,400	350,400
$R_{yrsw,SPST}$	1752	13,477	6147	7964
$R_{yrsw,SPDT}$				8760
$N_{yr,SPST}$	57	7	16	13
$N_{yr,SPDT}$				7

According to that, the total cost evaluation, including the overall system, is considered and reported in Table 7 for different cases of years to come before the switches are changed.

Table 7. Cost evaluation according to the estimated endurability, as reported in Table 6 in the four cases, for the "worst case" condition.

After n Years	Switches Cost Evaluation			
	Case 1	Case 2	Case 3	Case 4
n = 10	0	67.6	0	208.8
n = 20	0	135.2	148.2	532
n = 30	0	270.4	148.2	1064
After n Years	**Total Cost Evaluation**			
	Case 1	Case 2	Case 3	Case 4
n = 10	1 657 €	251 €	431 €	1 357 €
n = 20	1 657 €	319 €	579 €	1 680 €
n = 30	1 657 €	454 €	579 €	2 212 €

The economical contributions concerning the switches and the overall system, as reported in Table 7, arise from the data reported in Tables 1 and 2, respectively. Note that the configurations with the lowest number of switches are less convenient if only the price of the switches is considered, supposing that in the same number of years they require to be changed a higher number of times. On the contrary, if the total cost of the reconfigurator is considered, the cases with the lowest number of switches are the most convenient. Indeed, the initial price in terms of sensors, drivers and MOSFETs is generally higher if the number of mechanical switches is higher, due to the higher hardware complexity. Note as well as that if a low number of switches is associated to a more complex algorithm, so that the reconfiguration frequency is higher, the frequency of maintenance increases. As an example, Table 8 refers to two reconfigurations per minute in case 2, whereas the number of reconfigurations per minute in the other cases is kept at 1.

Table 8. Cost evaluation if in case 2 (case of minimum number of switches) the number of reconfigurations per minute is 2 instead of 1.

Total Cost for Years	Case 1	Case 2	Case 3	Case 4
Reconfigurations per minute	1	2	1	1
after n years		Total cost evaluation		
n = 10	1 657 €	319 €	431 €	1 357 €
n = 20	1 657 €	522 €	579 €	1 680 €
n = 30	1 657 €	725 €	579 €	2 212 €

One can see that in this case the most convenient solution, after 30 years, is the one corresponding to the case 3. Although the obtained economical results of this comparison among different reconfiguration cases shall not be critically considered, what is significant in this paragraph is the proposed approach for an economical estimation of the system lifetime.

3. Experimental Set-Up

Figure 5 sketches a realistic situation in which, during the day, a shadow overlays different panels. As a consequence, the shadowed panels disturb those connected in parallel: a decrease in voltage of shadowed panel involves a decrease of not shadowed ones and a rise of current for the power balance. This rise of current is not constantly probable: if the not shadowed panels are in the area of maximum power point, any decrease of voltage decreases the power.

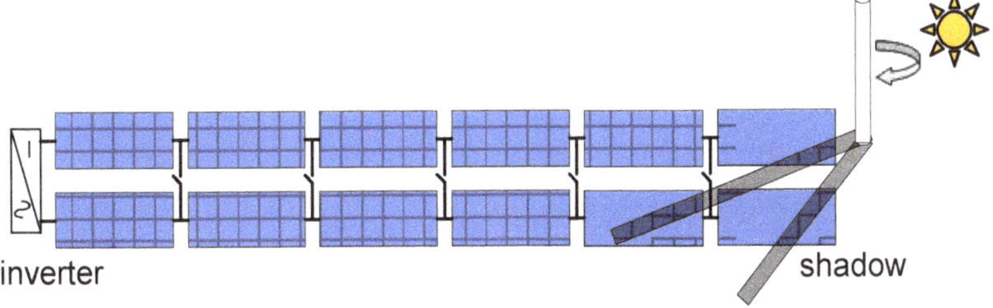

Figure 5. Probable shadow projections in the studied situation.

Figure 6 shows the effect of a shadow due to the presence of a pole by considering a PV DRS scheme. The hypothesis is that each panel has three lines of cells and is connected to the DRS. DRS is connected to a two channels inverter.

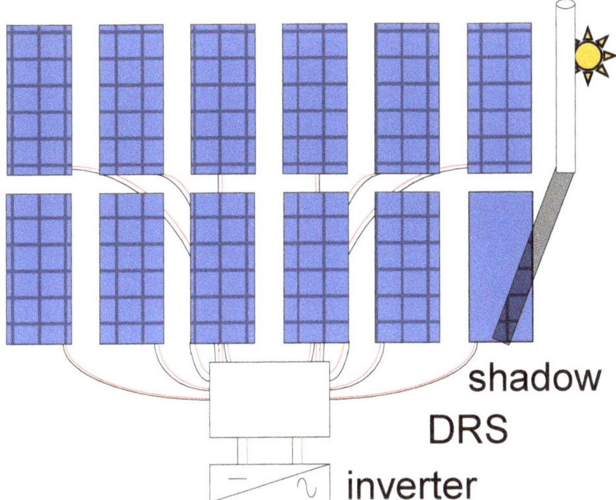

Figure 6. DRS in a PV plant.

Reconfiguration performances were tested with a prototype DRS developed at the University of Palermo and working on a twelve panels system. DRS acquires the state of every panel with a sensing system (voltage, current and temperature).

Figure 7 shows the experimental DRS system. Twelve panels (PMMP 215 W, VMPP 28.27 V, IMPP 7.59 A, Voc 36.37, Isc 8.21 A, Conergy, (Hamburg, Germany) have been connected individually to the DRS. Panels located at latitude: 38° 5.9′; longitude: 13° 20.6′, measurement carried out in the morning of 27 April 2016, from 10:00 to 13:00, with incident radiation between 1.0 and 1.1 kW/m^2 and temperature between 20–25 °C.

Without shadows, each panel works in the same point: the DRS generates basic topology, in this case, two parallel identical strings of six modules in series. To perform the behavior of the DRS, different resistive loads have been considered and Table 9 lists the different working conditions.

In order to test the DRS, an artificial shadow has been created. Figure 8 displays three shadow cases for a panel connected in series with five others. The artificial shadow cuts one, two or three lines of modules, dropping the performance of the panel and of the string. Each stoppage of line requires the action of the bypass diode, and a successive voltage decrease of the panel. The shadowed panel has labelled with the number 6, so V1–V5 are the voltages of normally irradiated panels and V6 is the shadowed one; DRS evaluates the power of each panel and connects the panels into the string.

Figure 7. Experimental system with panels individually connected to DRS.

Table 9. Electrical characteristics in different load conditions evaluated by reconfigurator for each panel.

Electrical Characteristics	Load A	Load B	Load C	Load D	Load E
Voltage (V)	31.3	28.9	26.1	23.1	16.8
Current (A)	1.9	4.0	6.0	7.0	7.4
Power (W)	54.5	115.6	156.6	161.7	124.3
String (W)	356.2	693.7	939.2	970.2	745.5

Figure 8. Shaded panel case 1, 2 and 3. In case 1 shadow can vary from 225 to 450 cm^2; which corresponds to one interruption of a line of cells; in case 2 shadow varies from 450 to 900 cm^2, which corresponds to two interruption of lines; case 3 corresponds to the interruption of the panel.

Table 10 recapitulates the behavior of the DRS with shading and without; P_1–P_5 are the powers of the not shadowed panels, P_6 is the power of the shadowed one in three cases, I is the current of the system in the different cases taken into account

Table 10. Electrical characteristics of the string in different shadow conditions.

Load	Shadow	V_{1-5} (V)	V_6 (V)	I (A)	P_1-P_5 (W)	P_6 (W)	P_{string} (W)
Load A	Not shaded	31.3	31.3	1.9	54.5	54.5	356.2
	case 1	31.6	20.2	1.6	50.5	32.3	285.1
	case 2	31.8	9.0	1.2	38.1	10.8	201.6
Load B	Not shaded	28.8	28.8	4.0	115.2	115.2	691.2
	case 1	29.2	18.7	3.3	96.6	61.7	543.5
	case 2	30.1	8.2	2.6	78.3	21.3	412.6
Load C	Not shaded	26.1	26.1	6.0	156.6	156.6	939.6
	case 1	27.7	17.1	5.0	138.5	85.5	778.0
	case 2	29.1	7.4	4.2	122.2	31.1	642.2
Load D	Not shaded	23.1	23.1	7.0	161.7	161.7	970.2
	case 1	25.9	15.6	6.4	165.7	99.8	928.6
	case 2	27.7	6.8	5.2	144.0	35.4	755.6
Load E	Not shaded	16.8	16.8	7.4	124.3	124.3	745.9
	case 1	16.8	11.0	7.4	124.3	81.4	703.0
	case 2	16.8	4.2	7.4	124.3	31.1	652.6

Figures 9 and 10 show respectively the voltage-current and voltage-power profiles of the panel with three working lines of cells, two working lines of cells and only one working line. Blue curves describe the working points without any shading. If a line of cells is shaded the orange curve has to be considered, maintaining a similar current and with a new voltage. When a shadow covers two lines of cells from the blue curve the gray curve has to be considered, maintaining a similar current and with a new voltage.

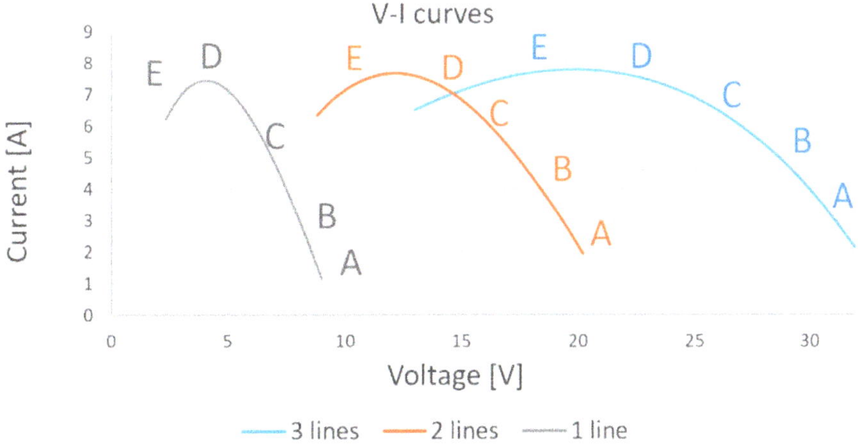

Figure 9. Interpolated V-I curves of the panel with different shadows. Working lines have been described.

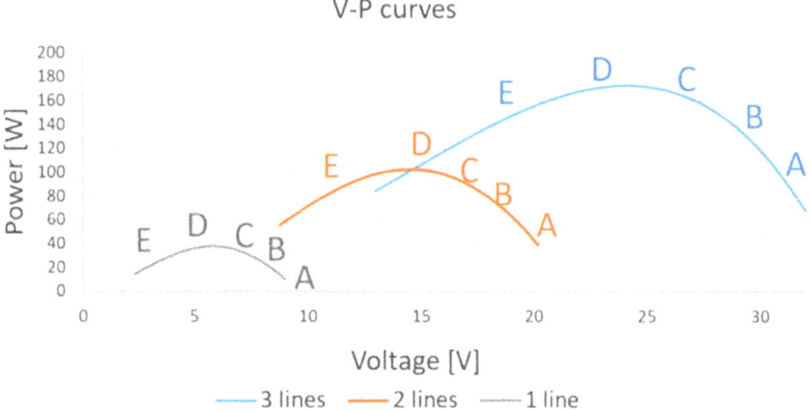

Figure 10. Interpolated V-P curves of the panel with different shadows. Working lines have been described.

New voltages indicate the new power conditions. The DRS is able to regroup similar irradiated panels and/or exclude the densely shaded panels. The different operation is due to the DRS architecture and the algorithm implemented. A not smart DRS can only exclude shaded panels, a high-performance DRS relocates them on suitable dynamic arrays.

3.1. Evaluation of Power Losses for a Single Shaded Panel

In the following part three cases are evaluated: case 1, only a line has been interrupted; case 2, two lines were interrupted; case 3, the whole panel is shadowed. By excluding the shaded panels there will be 16.7% of losses. DRS can enable the power increases shown in Table 11.

Table 11. Electrical Characteristics in different loading conditions evaluated by reconfigurator for each panel.

Electrical Losses	Loss$_{case1}$%	Loss$_{case2}$%	Reco (W)	Loss Rec%	ΔP1%	ΔP2%
Load A	20.0	43.4	297.3	16.7	+3.5	+26.9
Load B	21.4	40.3	576.0	16.7	+4.8	+23.7
Load C	17.2	31.6	783.0	16.7	+0.6	+15
Load D	4.3	22.2	808.5	16.7	-12.3	+5.6
Load E	5.7	12.5	621.6	16.7	-10.9	−4.1

Data presented in Table 11 can be plotted in Figure 11: in case 1 and for the lower currents (load A, B and C) there is an increase of power, for higher currents (loads D and E) there is a decrease; in case 2, for lower currents (loads A, B, C and D) there is an increase, for the higher current (load E) there is a decrease.

Case 3 is now considered: the shadow cuts entirely the panel, as shown in Figure 8. A negative voltage of the shadowed panel affects the performance of the string. Each not shaded panel varies its operating condition assuming a voltage slightly higher than the non-perturbed, to compensate the voltage drop on the shaded panel and to try to maintain a high current. Table 12 shows the increase of power when a panel is totally shaded. The study of case 3 shows that when the shadows cuts in two parts the panel 6 and it becomes a load, reconfiguration reduces always the loss of power.

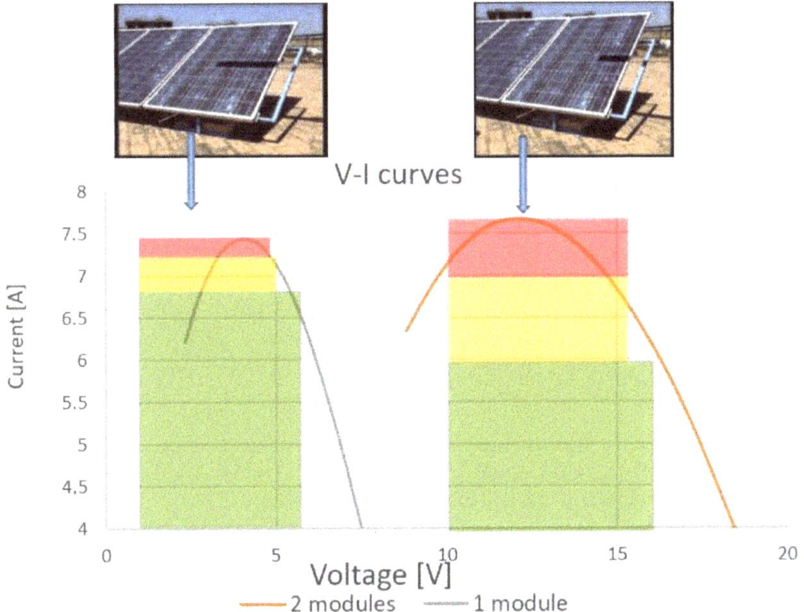

Figure 11. Zones of convenience of the disconnection of shaded module.

Table 12. Characteristics in different load conditions evaluated by reconfigurator for each panel.

Loads	Shadow	V_{1-5} (V)	V_6 (V)	I (A)	P_{1-55} (W)	P_6 (W)	P_{string} (W)	Loss%	ΔP3%
	Not shadow	26.1	26.1	6.0	156.6	156.6	939.6	-	-
Load C	Case 3	26.7	−2.9	5.1	136.1	−14.8	665.7	−29.1	-
	reconfigurated	26.1	open	6.0	156.6	-	783.0	-16.7	+12.4
	Not shadow	23.1	23.1	7.0	161.1	161.1	970.2	-	-
Load D	Case 3	23.7	−3.0	6.5	154.0	−19.5	750.7	−22.7	-
	reconfigurated	23.1	open	7.0	161.1	-	808.5	-16.7	+6.0
	Not shadow	16.8	16.8	7.4	124.3	124.3	745.9	-	-
Load E	Case 3	17.4	−3.1	7.4	128.7	−21.5	620.8	−16.7	-
	reconfigurated	16.8	open	7.4	156.6	-	621.5	−16.7	0

Case 3 shows the real performance of the DRS. Between loads C and D the maximum power point is performed. By considering the overall behavior of two shaded strings used simultaneously, a possible performance increase between 12 and 25% can be expected.

3.2. Evaluation of Power Losses for two shaded panels

The previously obtained performances only consider a shaded panel, and the logic is to maintain it (partially shaded) or exclude it (totally obscured). Now if two panels are shaded there is the possibility of placing in parallel. Panels named 6 and 7, can be re-configured to ensure the maximum current of the arrays. This feature requires a more evolved DRS, which can put in parallel panels belonging to different arrays, the sum of which currents is equal to the current of not obscured panels. This feature is useful only when it considers case 2 as shown in Table 13. The performances of different DRS are a mixture of the ones presented in Tables 12 and 13. For the economic analysis the increased power given by the reconfigurator is taken as an increase of the energy produced during the day.

Table 13. Electrical characteristic of the string in different shadow conditions.

Electrical Parameters	Load C			Load D		
	Not Shaded	Case 1	Case 2	Not Shaded	Case 1	Case 2
V_{1-5} (V)	26.1	27.7	29.1	23.1	25.9	27.7
V_{6-7} (V)	26.1	17.1	7.4	23.1	15.6	6.8
I (A)	6.0	5.0	4.2	7.0	6.4	5.2
P_{1-5} (W)	156.6	138.5	122.2	161.7	165.7	144.0
P_{6-7} (W)	156.6	85.5	31.1	161.7	99.8	35.4
P_{string} (W)	1096	863.5	673.3	1132	1028	755.6
Loss%	-	-21.2	-38.5	-	-9.2	−33.2
V_{6-7} (V)	-	19.3	6.2	-	15.0	5.0
I_{6-7} (A)	-	6.0	6.0	-	7.0	7.0
P_{6-7} (W)	156.6	60	36.1	161.7	101.0	36.3
P_{string} (W)	1096	903.3	909.2	970.2	1010	881.1
ΔP%	-	+4.2	+21.4	-	−1.5	+11.1

4. Economic Data

In order to carry out a complete study, the economic analysis presented in this paper takes into account different aspects of a PV plant, such as PV technology, location of the installation, government incentives, component lifetime, aging and periodic maintenance costs.

This study is based on the use of economic a tool called net present value (NPV), in order to evaluate the benefits of an economic investment in PV field with innovative devices such as a DRS. The NPV allows to evaluate the economic convenience of an investment for a specific period from a sum of cash flows actualized at time zero. In (6) the mathematic expression to evaluate the NPV is reported:

$$\text{NPV} = -C_0 + \sum_{y=1}^{n} \frac{C_y}{(1+i)^y} \tag{6}$$

With reference to expression (6), C_0 is the cash flow at time zero, C_y is the cash flow at the year y, i is the interest rate and y is the year of investment. Thus, the sign of the NPV, positive or negative, indicates the infeasibility of the investment and therefore the economic convenience. The interest rate i% considered in this work is equal to 5%. In this study, the technical-economic analysis has been carried out by considering an investment time equal to 20 years. In the following, detailed description of different aspects taken into account for the economic analysis and economic data are reported.

4.1. PV Plant

Since the first PV plants became commercially available on the market the cost of PV components, installation and maintenance has changed notably. In particular, in the last years there has been a constant decrease of the costs on the world market.

In this study, only residential PV systems with 6 kWp of power have been taken into account. This choice is motivated by the fact that the use of DRS is very interesting in residential PV plants, where the probability of installation of a fixed obstacle is high in respect to other type of plants. Indeed, four years after the installation, a fixed obstacle reducing by 35% the total power is assumed to appear. This value of reduction has been demonstrated in study [20]. Moreover, in order to carry out a complete analysis among different locations of installation taken into account, an installation of the PV plant between 2013 and 2014 was assumed, considering that in those years there were incentives in all the countries.

Regarding the cost of the PV plant, the average estimated price for this type of plant is about 2500 €/kWp (included installation). In Table 14 the economic data about the PV plant under test are summarized.

Table 14. PV plant data.

PV Plant Type	Residential Grid Connected
Power	6 kWp
Number of modules	20
PV plant Cost (installation included)	15, 000€
Power reduction	35%
Year of installation	2013/2014

4.2. Location of Installation and Economic Aspects

The installation location represents an interesting point of analysis in terms of production capability, government incentives and payments to private citizens for the production of the energy. Moreover, also the lifestyle of people plays an important role and therefore the average consumption per capita of the electric energy. In order to extend the economic analysis Italy, Germany, France, Spain, Bulgaria, Romania, Greece and Croatia have been considered as reference countries. As well known, these countries have different PV plant performances, different policies to improve the use of electricity generated from renewable sources and different people's lifestyle. In detail, the common strategy is based on a feed-in tariff (FIT) system with different values and timing of the incentives.

The economic data for each country (average consumption per capita, production facility, energy cost and incentives) have been referred of a PV plant installed in the capital of each countries. Moreover, has been considered a family composed by four people that lives in the capital of each countries. In Table 15 the considered economic data are reported.

Table 15. Economic data of the reference countries reproduced from [20], reproduced from proceedings of the 2018 International Conference on Smart Grid, IEEE: 2018.

Country	Average Consumption per Capita kWh/year)	Production Facility (kWh/year)	Energy Cost (€/kWh)	Incentives per Years (€/kWh)	Incentive Duration (years)
Italy	3200	9900	0.200	0.208	20
Germany	3512	6240	0.330	0.130	20
France	6343	7020	0.180	0.280	20
Spain	4131	9960	0.280	0.340	20
Bulgaria	4640	9000	0.090	0.240	20
Romania	2495	8400	0.125	0.160	15
Greece	5029	11,100	0.180	0.140	20
Croatia	3754	9600	0.132	0.150	14

These data are based on the following consideration:

- a family composed of four members and living in each capital of the considered countries;
- the electrical energy produced per year by a 6 kWp PV plant has been taken into account.
- the installation of the plant has been assumed between the year 2013 and 2014.

It should be noted that France has the highest value of average consumption per capita. This data is very high with respect to the production facility of the PV plant, so the analysis for the use of DRS is particularly difficult. The higher value of the energy production is in Greece, whereas, Romania and Croatia only offer incentives for 15 and 12 years. It should be noted that these considerations have influenced the economic results.

4.3. Inverter

The inverter represents the heart of the production from the solar energy. In particular, as well known, this system allows the electric energy conversion from DC to AC in order to inject the power surplus into the grid. Therefore, in the case of inverter fault it is not possible to use the energy with a consequence economic loss. From different studies, it is known that the inverter is the component more sensitive to failure. In [2], the authors considered an inverter life equal to 10 years. Nevertheless, it is not possible to estimate with accuracy the lifetime of an electronic component. For this reason, by considering a possible worst case, in this study it was assumed that the average lifetime of the inverter is equal to seven years. As far as the cost is concerned, according to [36] for a residential PV plant with 6 kWp of power the average cost is equal to 1000 €.

4.4. Increase of Production by DRS

The purpose of a DRS system is to increase the power of a PV plant in the case of a reduction of the total power production. The increment of the power production is a parameter that depends on the DRS topology and therefore hard to estimate. Indeed, the increment of power depends on the type of DRS and therefore of the number of possible available configuration. Thus, the number of possible reconfiguration available play an important role. By considering that a DRS with high number of switches allows many reconfigurations with respect to a DRS with a low number of switches, it is supposed to provide a higher increment of power. Nevertheless, this consideration is not enough to estimate the increment of power provided by a DRS with a defined number of switches, because it is possible that a DRS with high number of switches has redundant configurations. For this reason, in this work the same value of power increment has been considered for each DRS and fixed equal to 10% for the sake of simplicity. This obviously represents an unfavourable condition for DRS with high number of switches and higher costs. Nevertheless, this choice allows to emphasize the effect on the economic analysis of the costs and lifetime for each DRS.

4.5. Aging and Maintenance of PV Plant

After the installation, a natural phenomenon is the aging of the PV components. This phenomenon causes a reduction of the power and it increases over the time. Thus, in the economic analysis has been considered a reduction of power after the first year of the installation equal to 3% and a reduction for each year equal to 0.5%. Moreover, also a periodic maintenance has been considered with a cost equal to 100 €/year.

5. Economic Results and Discussion

As described above, the economic analysis of this study is focused on the evaluation of the economic benefits by using a DRS system four years after the installation of the PV plant with a power reduction equal to 35%. In particular, four cases of DRS have been analysed in different EU countries in order to extend the economic results. Figure 12 shows the NPV trend over the time of four cases of DRS_{1-4} and without DRS for each country.

The best result has been obtained in Spain with the highest value of the NPV after 20 years due to the incentives per year. Positive NPV values have been obtained in Italy, Bulgaria and Greece thanks to the high values of the production facility. Romania and Croatia have been penalized for a lower duration of the incentives, whereas, France and Germany have been penalized for the lower values of the production facility. The values of the NPV for each country and for each DRS are summarized in Table 16.

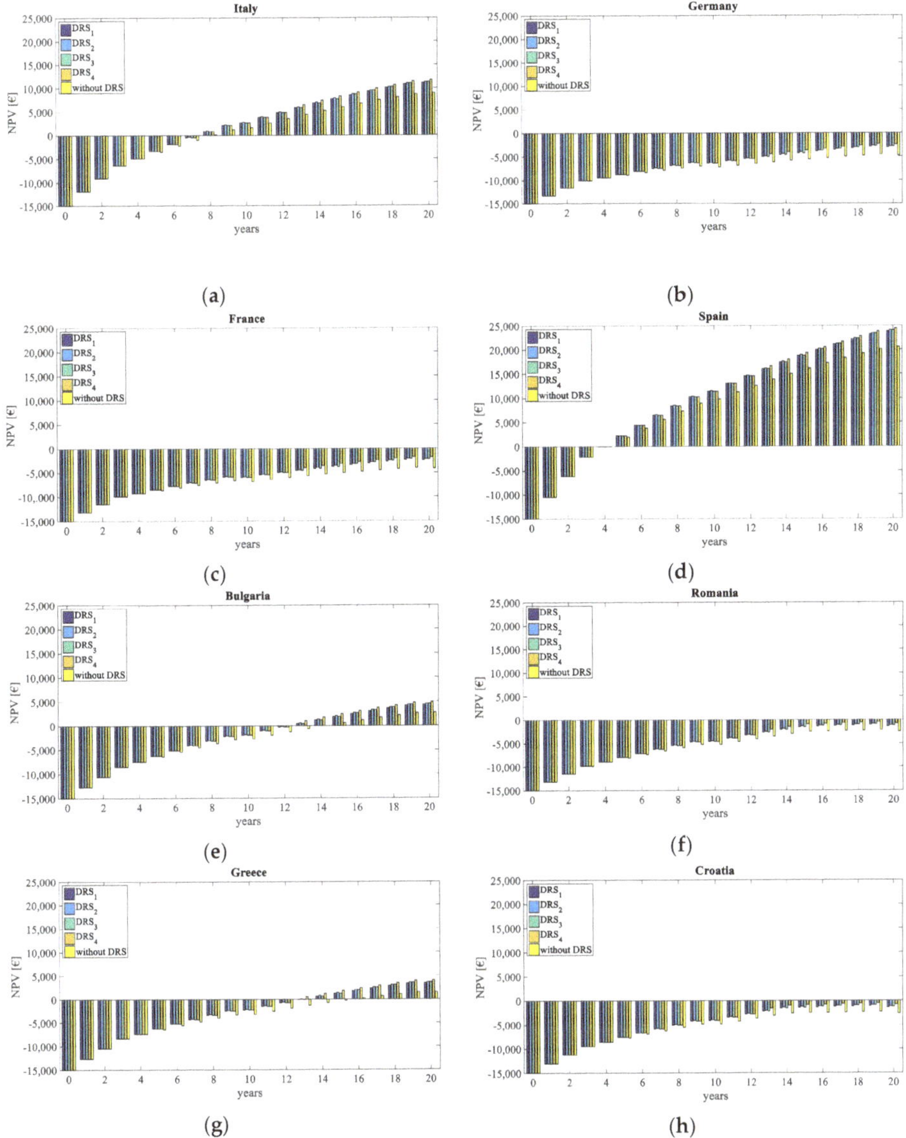

Figure 12. NPV trend over the time of four cases of DRS_{1-4} and without DRS, in (**a**) Italy, (**b**) Germany, (**c**) France, (**d**) Spain, (**e**) Bulgaria, (**f**) Romania, (**g**) Greece and (**h**) Croatia.

By analysing the NPV values of Table 16, it is interesting to note that DRS_4 allows one to obtain the best results also in the cases in which there the negative values of NPV. Moreover, this result is interesting because the DRS_4 present the second highest cost equal to 1148 € also by considering the worst case for the DRS with higher costs. Other interesting consideration can be done by changing the increment power. By considering an increment of the power equal to 20%, the NPV values obtained are reported in Table 17.

Table 16. NPV values after 20 years (increment power 10%).

Countries	NPV After 20 Years (€)				
	DRS$_1$	DRS$_2$	DRS$_3$	DRS$_4$	Without
Italy	10,871	11,095	11,068	11,480	8723
Germany	−3298	−3075	−3101	−2689	−4991
France	−2663	−2439	−2466	−2054	−4380
Spain	23,633	23,856	23,830	24,241	20,349
Bulgaria	4076	4300	4274	4685	2497
Romania	−1601	−1377	−1403	−992	−2713
Greece	3186	3410	3384	3795	1297
Croatia	−1646	−1423	−1449	−1038	−2873

Table 17. NPV values after 20 years (increment power 20%).

Countries	NPV After 20 Years (€)				
	DRS$_1$	DRS$_2$	DRS$_3$	DRS$_4$	Without
Italy	13,019	13,243	13,217	13,628	8723
Germany	−1606	−1382	−1408	−997	−4991
France	−946	−722	−748	−337	−4380
Spain	26,917	27,140	27,114	27,526	20,349
Bulgaria	5656	5879	5853	6265	2497
Romania	−489	−265	−291	120	−2713
Greece	5075	5299	5273	5684	1297
Croatia	−420	−196	−222	189	−2873

In respect to the previous case, the DRS$_4$ with an increment of power equal to 20% allows to obtain positive values of the NPV also for Romania and Croatia. This result is realistic because the DRS$_4$, thanks to the high number of switches and thus the high number of the possible configurations, may generate an increment of power equal to 20%. Another interesting point of analysis is the payback time. In Table 18, the payback times for each country and for each DRS in two power increment cases are reported.

Table 18. Payback time for increment of power equal to 10% and 20%. In some cases the payback time exceeds the reasonable time for the return on the investment, and it is not evaluated (n.e.).

Countries	Payback Time (Years)							
	DRS$_1$		DRS$_2$		DRS$_3$		DRS$_4$	
	10%	20%	10%	20%	10%	20%	10%	20%
Italy	8	8	8	8	8	8	8	8
Germany	n.e.	28	n.e.	27	n.e.	27	n.e	25
France	n.e.	25	n.e.	24	n.e.	24	n.e.	23
Spain	5	5	5	5	5	5	5	5
Bulgaria	13	12	13	12	13	12	13	12
Romania	n.e.	n.e.	n.e.	27	n.e.	n.e.	n.e.	20
Greece	15	12	14	12	15	12	13	12
Croatia	n.e.	n.e.	n.e.	n.e.	n.e.	n.e.	n.e	19

Spain and Italy present the best results and it is interesting to note that the same values in the two cases of the increment of power have been obtained. In other countries (Germany, France and Romania), it was necessary to extend the duration of the investment in order to find the payback time but in all cases the best results have been obtained with DRS$_4$ and an increment of power equal to 20%.

Only in Bulgaria and Greece a reduction of the payback time has been obtained, an increment power equal to 20%, 13 years to 12 years and from 15 years to 12 years, respectively.

6. Conclusions

This paper presents a complete analysis on the real benefits introduced by a DRS system in a PV plant after a considerable power reduction. In particular, in order to evaluate the overall economic impact on the costs of a PV plant the technical and economic aspects about the DRS have been considered. In the first part of the paper an economic analysis of different DRSs due to the costs of the components and to the adopted topological schemes, is carried out. Switching matrix, sensing network and driving circuit constitute the architecture of DRS, the choice of switches and their number affects the electrical endurance. A more flexible DRS involves higher initial cost due to the number of switches required by the adopted architecture, but at the same time a less exploitation and therefore a longer useful life.

From the economic point of view, the analysis has been extended to different countries of EU in order to considerate incentives policies, location of installation and lifestyles of the people, while, from a technical study, that takes into account the hardware complexity in terms of the components required and other technical aspects, the costs and lifetime of four DRS have been estimated. The economic tool used in this analysis is the NPV and the payback time.

Firstly, in all scenarios the analysis has demonstrated the positive economic impact on the use of a DRS in a PV plant with respect to the cases without DRS. The best results have been obtained in Spain thanks to the higher value of the incentives per year in terms of NPV and payback time, whereas good results have been obtained for Italy, Bulgaria and Greece. The worst results have been obtained for France and Germany due to the lower values of the production facility. Among the DRS considered in the economic analysis, the DRS_4 option allows one to obtain the best results in all cases. The best performance of DRS_4 is attributable to the high number of switches that allows it to increase the lifetime of the system.

Author Contributions: Conceptualization, F.V.; methodology, F.V. and P.R.; software, G.S. and F.P.; validation, F.P. and G.S.; formal analysis, F.V. and G.S.; investigation, P.R. and F.V.; resources, R.M. and G.A.; data curation, G.S. and F.P.; writing—original draft preparation, F.P., G.S., and F.V.; writing—review and editing, G.A. and F.V.; visualization, G.A. and R.M.; supervision, G.A., R.M. and F.V. All authors have read and agreed to the published version of the manuscript.

Funding: This research received no external funding. This work was financially supported by MIUR-Ministero dell'Istruzione, dell'Università e della Ricerca (Italian Ministry of Education, University and Research) and by SDESLab (Sustainable Development and Energy Saving Laboratory) and LEAP (Laboratory of Electrical APplications) of the University of Palermo.

Conflicts of Interest: The authors declare no conflict of interest.

References

1. Timilsina, G.R.; Kurdgelashvili, L.; Narbel, P.A. Solar energy: Markets, economics and policies. *Renew. Sustain. Energy Rev.* **2012**, *16*, 449–465. [CrossRef]
2. Branker, K.; Pathak, M.J.M.; Pearce, J.M. A review of solar photovoltaic levelized cost of electricity. *Renew. Sustain. Energy Rev.* **2011**, *15*, 4470–4482. [CrossRef]
3. Lappalainen, K.; Valkealahti, S. Photovoltaic mismatch losses caused by moving clouds. *Sol. Energy* **2017**, *158*, 455–461. [CrossRef]
4. Zou, X.; Li, B.; Zhai, Y.; Liu, H. Performance monitoring and test system for GridConnected photovoltaic systems. In Proceedings of the 2012 Asia-Pacific Power and Energy Engineering Conference, Shanghai, China, 27–29 March 2012.
5. Caruso, M.; Miceli, R.; Romano, P.; Schettino, G.; Spataro, C.; Viola, F. A low cost, Real-time monitoring system for PV plants based on ATmega 328P-PU Microcontroller. In Proceedings of the 2015 IEEE International Telecommunications Energy Conference (INTELEC), Osaka, Japan, 18–22 October 2015.

6. Caruso, M.; Di Tommaso, A.O.; Miceli, R.; Ricco Galluzzo, G.; Romano, P.; Schettino, G.; Viola, F. Design and experimental characterization of a low cost, real-time, wireless, AC monitoring system based on ATmega 328P-PU microcontroller. In Proceedings of the 2015 AEIT International Annual Conference (AEIT), Naples, Italy, 14–16 October 2015; pp. 1–6.
7. Wang, Y.-J.; Hsu, P.-C. An investigation on partial shading of PV modules with different connection configurations of PV cells. *Energy* **2011**, *36*, 3069–3078. [CrossRef]
8. Potnuru, S.R.; Pattabiraman, D.; Ganesan, S.I.; Chilakapati, N. Positioning of PV panels for reduction in line losses and mismatch losses in PV array. *Renew. Energy* **2015**, *78*, 264–275. [CrossRef]
9. Wang, Y.J.; Lin, S.S. Analysis of a partially shaded PV array considering different module connection schemes and effects of bypass diodes. In Proceedings of the 2011 International Conference & Utility Exhibition on Power and Energy Systems: Issues and Prospects for Asia (ICUE), Pattaya City, Thailand, 28–30 September 2011; pp. 1–7.
10. Kandemir, E.; Cetin, N.S.; Borekci, S. A comprehensive overview of maximum power extraction methods for PV systems. *Renew. Sustain. Energy Rev.* **2017**, *78*, 93–112. [CrossRef]
11. Kouchaki, A.; Iman-Eini, H.; Asaei, B. A new maximum power point tracking strategy for PV arrays under uniform and non-uniform insolation conditions. *Sol. Energy* **2013**, *91*, 221–232. [CrossRef]
12. Alahmad, M.; Chaaban, M.A.; Kit Lau, S.; Shi, J.; Neal, J. An adaptive utility interactive photovoltaic system based on a flexible switch matrix to optimize performance in real-time. *Sol. Energy* **2012**, *86*, 951–963. [CrossRef]
13. Ramaprabha, R.; Mathur, B.L. A Comprehensive review and analysis of solar photovoltaic array configurations under partial shaded conditions. *Int. J. Photoenergy* **2012**, *2012*, 1–16. [CrossRef]
14. Candela, R.; Sanseverino, E.R.; Romano, P.; Cardinale, M.; Musso, D. A Dynamic electrical scheme for the optimal reconfiguration of PV modules under non-homogeneous solar irradiation. *Appl. Mech. Mater.* **2012**, *197*, 768–777. [CrossRef]
15. Sanseverino, E.R.; Ngoc, T.N.; Cardinale, M.; Romano, P.; LiVigni, V.; Musso, D.; Viola, F. Dynamic programming and Munkres algorithm for optimal photovoltaic arrays reconfiguration. *Sol. Energy* **2015**, *122*, 347–358. [CrossRef]
16. Tian, H.; Mancilla–David, F.; Ellis, K.; Muljadi, K.; Jenkins, P. Determination of the optimal configuration for a photovoltaic array depending on the shading condition. *Sol. Energy* **2013**, *95*, 1–12. [CrossRef]
17. Balato, M.; Costanzo, L.; Vitelli, M. Series-parallel PV array re-configuration: Maximization of the extraction of energy and much more. *Appl. Energy* **2015**, *159*, 145–160. [CrossRef]
18. Chao, K.H.; Ho, S.H.; Wang, M.H. Modeling and fault diagnosis of a photovoltaic system. *Electr. Power Syst. Res.* **2008**, *78*, 97–105. [CrossRef]
19. Livreri, P.; Caruso, M.; Castiglia, V.; Pellitteri, F.; Schettino, G. Dynamic reconfiguration of electrical connections for partially shaded PV modules: Technical and economical performances of an Arduino-based prototype. *Int. J. Renew. Energy Res.* **2018**, *8*, 336–344.
20. Viola, F.; Romano, P.; Miceli, R.; Spataro, C.; Schettino, G. Technical and Economical Evaluation on the Use of Reconfiguration Systems in Some EU Countries for PV Plants. *IEEE Trans. Ind. Appl.* **2017**, *53*, 1308–1315. [CrossRef]
21. Caruso, M.; Miceli, R.; Romano, P.; Sanseverino, E.R.; Schettino, G.; Viola, F.; Ngoc, T.N. Comparison between different dynamic reconfigurations of electrical connections for partially shaded PV modules. In Proceedings of the 2018 International Conference on Smart Grid (icSmartGrid), Nagasaki, Japan, 4–6 December 2018; pp. 220–227.
22. Iraji, F.; Farjah, E.; Ghanbari, T. Optimisation method to find the best switch set topology for reconfiguration of photovoltaic panels. *IET Renew. Power Gener.* **2018**, *12*, 374–379. [CrossRef]
23. Tabanjat, A.; Becherif, M.; Hissel, D. Reconfiguration solution for shaded PV panels using switching control. *Renew. Energy* **2015**, *82*, 4–13. [CrossRef]
24. Tria, L.A.R.; Escoto, M.T.; Odulio, C.M.F. Photovoltaic array reconfiguration for maximum power transfer. In Proceedings of the TENCON 2009—2009 IEEE Region 10 Conference, Singapore, 23–26 January 2009; pp. 1–6.
25. Chaaban, M.A.; Alahmad, M.; Neal, J.; Shi, J.; Berryman, C.; Cho, Y.; Lau, S.; Li, H.; Schwer, A.; Shen, Z.; et al. Adaptive photovoltaic system. In Proceedings of the IECON 2010—36th Annual Conference on IEEE Industrial Electronics, Glendale, AZ, USA, 7–10 November 2010; pp. 3192–3197.
26. RS Homepage. Available online: https://www.rs-online.com/ (accessed on 2 May 2019).

27. Balato, M.; Vitelli, M.; Femia, N.; Petrone, G.; Spagnuolo, G. Factors limiting the efficiency of DMPPT in PV applications. In Proceedings of the 2011 International Conference on Clean Electrical Power (ICCEP), Ischia, Italy, 14–16 June 2011; pp. 604–608.

28. Spagnuolo, G.; Petrone, G.; Lehman, B.; Paja, C.A.R.; Zhao, Y.; Gutierrez, M.L.O. Control of photovoltaic arrays: Dynamical reconfiguration for fighting mismatched conditions and meeting load requests. *IEEE Ind. Electron. Mag.* **2015**, *9*, 62–76. [CrossRef]

29. HF41F. Subminiature Power Relay. Available online: https://docs-emea.rs-online.com/webdocs/14c6/0900766b814c652d.pdf (accessed on 2 May 2019).

30. Finder. Ultra-Slim PCB Relays (EMR or SSR) 0.1-0.2-2-6 A. Available online: https://docs-emea.rs-online.com/webdocs/16d4/0900766b816d4f36.pdf (accessed on 10 February 2020).

31. Infineon. OptiMOS®2 Power-Transistor. Available online: https://datasheet.octopart.com/IPP08CN10N-G-Infineon-datasheet-5315235.pdf (accessed on 17 February 2020).

32. MAXIM. Isolated Transformer Driver for PCMCIA Applications. Available online: https://datasheets.maximintegrated.com/en/ds/MAX845.pdf (accessed on 1 March 2020).

33. Texas Instruments. MAX 845. LM35 Precision Centigrade Temperature Sensors. Available online: http://www.ti.com/lit/ds/symlink/lm35.pdf (accessed on 5 March 2020).

34. LEM. Current Transducer LTS 25-NP. Available online: https://docs-emea.rs-online.com/webdocs/146d/0900766b8146d120.pdf (accessed on 7 March 2020).

35. La Manna, D.; Vigni, V.L.; Sanseverino, E.R.; Di Dio, V.; Romano, P. Reconfigurable electrical interconnection strategies for photovoltaic arrays: A review. *Renew. Sustain. Energy Rev.* **2014**, *33*, 412–426. [CrossRef]

36. Fu, R.; Feldman, D.; Margolis, R.; Woodhouse, M.; Ardani, K. *U.S. Solar Photovoltaic System Cost Benchmark: Q1 2017*; Technical Report No. NREL/TP-6A20-68925 7752; USDOE Office of Energy Efficiency and Renewable Energy (EERE): Washington, DC, USA, 2017. [CrossRef]

Article

Reactive Power Injection to Mitigate Frequency Transients Using Grid Connected PV Systems

Yujia Huo, Simone Barcellona, Luigi Piegari and Giambattista Gruosso *

Politecnico di Milano, Dipartimento di Elettronica Informazione e Bioingegneria, Piazza Leonardo da Vinci 32, 20133 Milano, Italy; yujia.huo@polimi.it (Y.H.); simone.barcellona@polimi.it (S.B.); luigi.piegari@polimi.it (L.P.)
* Correspondence: giambattista.gruosso@polimi.it

Received: 22 January 2020; Accepted: 12 April 2020; Published: 17 April 2020

Abstract: The increasing integration of renewable energies reduces the inertia of power systems and thus adds stiffness to grid dynamics. For this reason, methods to obtain virtual inertia have been proposed to imitate mechanical behavior of rotating generators, but, usually, these methods rely on extra power reserves. In this paper, a novel ancillary service is proposed to alleviate frequency transients by smoothing the electromagnetic torque of synchronous generators due to change of active power consumed by loads. Being implemented by grid-tied inverters of renewables, the ancillary service regulates the reactive power flow in response to frequency transients, thereby demanding no additional power reserves and having little impact on renewables' active power generation. Differently from the active power compensation by virtual inertia methods, it aims to low-pass filter the transients of the active power required to synchronous generators. The proposed ancillary service is firstly verified in simulation in comparison with the virtual inertia method, and afterwards tested on processor by controller-hardware-in-the-loop simulation, analysing practical issues and providing indications for making the algorithm suitable in real implementation. The ancillary service proves effective in damping frequency transients and appropriate to be used in grid with distributed power generators.

Keywords: ancillary service; PV Plant; frequency-assisting; hardware-in-the-loop; Photovoltaic; DER

---

1. Introduction

Renewable energy sources (RESs) are of strategic importance and crucial to the sustainability of energy production from an environmental point of view. They are playing an important role in distribution networks [1,2] transforming them from passive to active [3], and representing a challenge for power planning and operation [4,5]. The increasing penetration of RESs, which are usually interfaced with static power converters, reduces the inertia of the electric network, so issues related to stiff frequency transients and oscillations arise and the immunity against faults and disturbances is weakened [6,7]. Nevertheless, RESs themselves are foreseen to play a major and decisive role in maintaining the reliability and stability of the grid in the future [8]. Moreover, the idea to divide the distribution networks in several small zones capable of self-regulating can help to overcome those issues. These kinds of partitioned networks, which are called microgrids, are well reviewed and classified in [9]. The photovoltaic (PV) system is one of the most representative RESs, which takes an active part in the construction of these modern grids. Therefore, the grid codes of many countries have requested for PV systems' cooperation in case of faults and transients [10,11]. Another important aspect is related to the management of both active and reactive power among different sources and loads. This aspect is crucial for microgrids that can operate both in grid connected mode and in islanded mode [12]. In any case, among the most popular control strategies, which also help for the stability of the network, there are the well-known droop-based controls. These kinds of control are very powerful because they can deal with different sources; they are flexible and can be used both in high voltage (HV) lines and low voltage (LV) distribution lines. In HV lines, it is well-know the assumption for which it is possible to decouple the frequency and voltage controls by acting on the active power (P/f) and reactive power (Q/V), respectively. This assumption is valid because in HV lines the inductive effect is predominant on the resistive one. On the contrary, in LV lines, where the resistive effect is the predominant one, the two droop controls can be exchanged each other having the P/V control and Q/f control [13–16]. In general, both effects are coupled. In particular, in the medium voltage lines the inductive and resistive effects are comparable, therefore, it is not possible none of the two assumptions. In this case, to decouple the problem it is possible to use the orthogonal linear rotational transformation matrix from the active and reactive actual powers to the modified ones as reported in [17]. Another way to decouple the two droop-controls, also considering the harmonic current in case of non linear loads, is through the usage of the virtual impedance [18–20].

Focusing, in particular, on the frequency stability of networks in which it is valid the assumption of predominant inductive effect, several approaches and control techniques have been studied and proposed. They can be categorized into two main groups: active power based frequency controls and reactive power based frequency controls. The active power based frequency controls are the most common used because they are related to the assumption itself of P/f and Q/V decoupling controls [21]. In this case, static reserves such as battery storage units can be dispatched to provide fast response when the system is under serious or peak load conditions [22,23]. Frequency-oriented advanced converter control algorithms, such as the virtual synchronous generator (VSG), are also viable methods tackling those problems [4,24,25]. They can be either provided by specific equipment or integrated into RES systems. In the former case, the energy reserve is utilized as a backup so it does not contribute to the nominal capacity. In the latter case, the availability of the VSG service could be restricted by ambient conditions; therefore, in order to ensure the service in any condition, auxiliary battery energy storage systems can be added to the RES [26]. The presence of additional equipment implies increased budget and complexity. In [27], the frequency-active power curve is studied and integrated into PV systems focusing on the curtailment, dead band and droop. Again, the droop-like control is limited by solar radiation and the efficiency of the PV systems is reduced. Ochoa [28] proposes an innovative control for wind system by shifting the maximum power point tracking (MPPT) to an optimized power point tracking, so that the power generation curve versus the ambient condition is smoothed. In this way, the wind farm contributes to the power grid capacity. Moreover, due to its self-regulating feature, it is not sensitive to grid conditions. All these methods present a common drawback related to the necessity to have an energy reserve to be implemented. For this reason, even if the active power based frequency controls are the most studied for the above-mentioned reasons, also the reactive power based frequency controls have been analyzed as well. It is worth nothing that, even if we are in the case in which it is valid the P/f and Q/V controls for the steady state power management, it is possible to use the reactive power during the transients to dump and help the frequency stability [29–35]. The first developed tools were the so-called power system stabilizers (PSSs) that act directly on the exciters of the synchronous generators (SGs) damping both the local and inter-area frequency oscillations [29,35]. Recently, thanks to the wide spread of the well-kwon flexible alternating current transmission systems (FACTSs) the supplementary damping controllers (SDCs) have also been developed [30–35]. Both the PSSs and SDCs provide an additional voltage signal to the voltage reference of the exciters of SGs or FACTSs. This signal is related to the frequency deviation. In order to avoid the influence of the voltage reference signal at the steady state regulation a washout filter is usually employed [34]. These methods, acting on the exciters of synchronous generators or on voltage references of FACTS are usually not really fast and, mainly, can be implemented only in few points of the grid.

Unlike the aforementioned previous researches, in this paper it is proposed a method based on a simple reactive power based frequency control carried out by RESs without introducing any additional reserve or reducing the active power efficiency. Moreover, no additional voltage reference is employed.

For this reason, the proposed method can be applied in any RES converter, spreading the service in the grid and improving, in this way, its effectiveness.

Taking the PV system as an example, its excessive capacity design makes it possible to generate or absorb a considerable amount of reactive power also at the peak hours. The capacity of the PV inverter is defined according to the maximum evaluated solar irradiance of the day. However, the inverter is not used at its maximum power during the whole day. Indeed, it stands partially idling in most of the daytime and completely idling after the sunset. Therefore, the active power produced by the PV can be controlled according to MPPT while exchanging reactive power. For this reason, the proposed control, based on exchanging reactive power to smooth frequency transients can be implemented without affecting the main PV control (i.e., MPPT). In the following, the proposed ancillary service will be shortly indicated as *Q/f* control.

In order to evaluate the effectiveness of the proposed ancillary service, it is firstly tested by means of numerical simulations. Frequency transients and the consumed power for this service are illustrated. For proving the hypothesis based on which the ancillary service is proposed, significant internal variables of the synchronous generator (SG) are shown as well. Moreover, a comparison between the proposed method and the VSG method is carried out focusing on the working principles, performance, cost effectiveness and practicability. Then a controller-hardware-in-the-loop (CHIL) platform is set up based on which, the ancillary service is tested in real time.

The paper is organized as follows: Section 2 reports the theoretical aspects of the proposed ancillary service; Section 3 demonstrates the effectiveness of the ancillary service in simulation and the comparison with the VSG method; Section 4 describes the performance of the algorithm in CHIL highlighting the related issues and suggesting appropriate solutions; finally Section 5 draws the conclusions.

2. The Proposed Ancillary Service

2.1. Scope and Comparison with VSG

Advanced control strategies, such as virtual inertia generation, are studied and implemented on grid-tied inverters to mitigate the stiffness caused by lack of inertia technologies. Virtual synchronous generators usually are implemented using a reference for the active power dependent on the virtual inertia, the virtual friction factor, the frequency and the rate of the frequency change (ROCOF) while the set point of the reactive power is obtained by the voltage regulator. This method partially compensates the fast active power variations of the loads during transients and thus gives less pressure to SG's governor to recover the frequency. The differential component ROCOF makes the active power loop respond to the frequency change earlier; however, as a consequence, it can bring instability concerns. VSG can be carried out by the highly integrated RESs. RESs are expected to work at the MPPs which are determined by ambient conditions and therefore, the MPPs can be taken as the output active power of VSG at steady state. When frequency goes below the setpoint, VSG is required to provide fast active power response greater than the MPP and the amount over MPP must be supported by other reserves. These reserves serve the transients but do not contribute to the capacity of the network.

The purpose of the proposed ancillary service is to alleviate frequency transients without either introducing additional power reserves or sacrificing the maximum active power generation. This is achieved, in this paper, by relating the grid frequency to the reactive power Q exchanged by the RES inverter with the grid itself.

The main idea is to change the voltage across the terminals of the SG with a consequent variation of the electromagnetic torque developed by the machine. In order to achieve this objective, the reactive power can be utilized. Indeed, while the excitation system fixes the rotor flux, controlling the reactive power it is possible to change the stator flux and, consequently, the machine torque. Acting on the torque, it is possible to modify the speed transient of the SG and, as a consequence, the grid frequency transient. Therefore, a relationship between grid frequency and reactive power can be implemented in the inverter control for a transient assisting purpose. Frequency transients are due to unbalance between prime motor torque and electromagnetic torque. When electrical load increases, the electromagnetic torque increases and, while the prime motor adjust its torque, the machine speed decreases and so does the frequency. For this reason, when, for example, the grid frequency drops below its set point, the proposed ancillary service makes the inverter to absorb reactive power reducing the stator flux and limiting the increasing of the electromagnetic torque mitigating the speed transient. Finally, the proposed service is implemented by means of a linear relationship between the reactive power absorbed by the inverter and the frequency deviation from the rated value.

Table 1 summarizes and compares the features of the VSG method and the proposed *Q/f* method. J is the virtual moment of inertia; F is the virtual friction factor; ω is the electrical angular frequency; v is the voltage; p is the number of pole pairs. P^* and Q^* are the active and reactive power set points of the ancillary service control loops. In the VSG method, the fast change of the load power is compensated by a third entity, i.e., the VSG.

Table 1. Features of VSG and the proposed Q/f method.

	SG Model Based VSG	Proposed Method
working principle	$P^* = f(J, F, \frac{\omega}{p}, \frac{d(\omega/p)}{dt})$ $Q^* = 0$ or $Q^* = f(v)$	$P^* = MPP$ $Q^* = f(\omega)$
effective power	P or P, Q	Q
extra reserve?	Yes	No
expected effect	compensating load power	impeding voltage recovery
other features	responds to $d\omega/dt$, early instability concerns	responds to ω, later high loop speed required

In order to achieve a low-pass filtering effect on the change of the electromagnetic torque, the proposed method must be faster than the governing and excitation systems of the SG. Mostly, both systems are much slower than the power electronic devices.

It is worth noting that, in low voltage distribution grids with low X/R ratio it is not possible to decouple the frequency and voltage regulation using active and reactive power respectively. In those cases, a cross-coupled regulation of active and reactive power is used to regulate frequency and voltage [36]. Nevertheless, for grid connected PV plants, active power is usually injected to support microgrids during frequency transients [37]. Moreover, it has to be highlighted that, the proposed service is a *fast-transient* service. Indeed, the regulation proposed has its positive effect in the fast transient when, usually, excitation systems of synchronous generators are not yet capable of working. A sort of secondary regulation leading the reactive power to zero after the frequency transient can be designed. In this paper, for sake of simplicity, this possibility has not been investigated.

2.2. Working Principle

Since the electrical frequency is set by the grid former which is usually a SG, the analysis is started from the mechanical behavior of the SG. Under the assumption of the friction absence the rotation motion of SG's rotor can be expressed by:

$$J \frac{d\omega}{dt} = T_m - T_e, \tag{1}$$

where T_m is the mechanical torque produced by the prime mover and T_e is the electromagnetic torque applied to the rotor. From the Equation (1) it can be easily understood that the frequency transient depends on the ratio between the torque difference and the inertia. Given a SG, the inertia is fixed and the mechanical torque is provided by a governor with a slow response. Therefore, during a load change which induces an abrupt variation of the electromagnetic torque, the frequency experiences either an over-shooting or a drop in addition to a long recovery process usually accompanied by oscillations due to the low speed of the governor's response.

The working principle of VSG is to compensate the slow response of the mechanical system by injecting or absorbing active power to the grid. The motion equation of SG's rotor is then changed into:

$$J\frac{d\omega}{dt} = T_m + \frac{\Delta P_{VSG}}{\omega} - Te = (T_m + \Delta T_m) - T_e, \tag{2}$$

where ΔP_{VSG} is the compensating power injected by the VSG. This direct compensation is achieved by active power control and the compensated power is then transformed into compensated torque. Consequently, there are three origins of torques influencing the motion of the rotor: the prime mover torque, T_m, the virtual mechanical torque ΔT_m resulted from the active power of VSG and the electromagnetic torque T_e given by electric loads.

Differently from VSG, the proposed method aims at damping the frequency oscillations during the transients by means of regulating the reactive power. The motion equation of SG's rotor is thus changed into:

$$J\frac{d\omega}{dt} = T_m - \frac{f_{LPF}(P_{load})}{\omega} = T_m - f_{LPF}(T_e), \tag{3}$$

where the function $f_{LPF}()$ stands for low-pass filtering behaviour. The high-frequency attenuation is achieved by additional reactive power absorption or injection which smooths the change of electromagnetic torque.

Figure 1 shows the dynamic equivalent model of SG in the dq rotating frame. All the parameters have been transformed to the stator side. For clarity, the losses and the presence of dampers are ignored. Subscripts d and q respectively represent the variables or parameters on d-axis and q-axis; Superscript ^{r} refers to the dq-frame; V_f and i_f are the excitation voltage and current; v is the terminal voltage of SG; i is the armature current; ϕ is the stator flux; L_l is the armature leakage inductance; L_m is the magnetizing inductance and L_{fl} is the field winding leakage inductance.

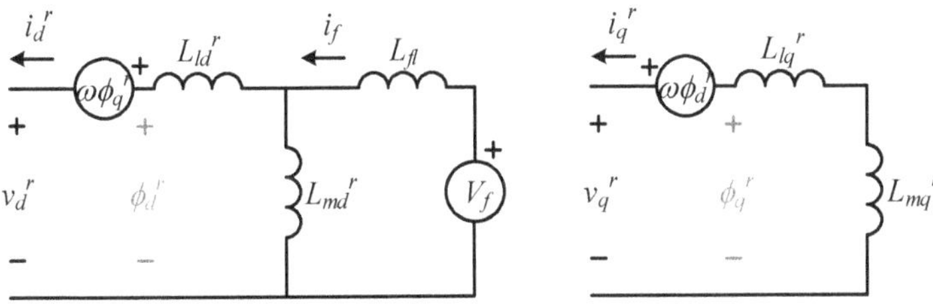

Figure 1. Equivalent dynamic model of SG seen at stator side in the dq frame set on the rotor.

As is known, the dq terminal voltages of SG are function of the stator flux, of its derivative and of the speed of the reference frame coinciding, at steady state, with the angular frequency. This is expressed, under lossless condition, as:

$$\begin{cases} v_d^r = \frac{d\phi_d^r}{dt} - \omega\phi_q^r \\ v_q^r = \frac{d\phi_q^r}{dt} + \omega\phi_d^r \end{cases}. \tag{4}$$

Therefore, in symmetric situation, the instantaneous active electric power of SG can be derived as:

$$P_e = \frac{3}{2}v_d^r i_d^r + \frac{3}{2}v_q^r i_q^r = \frac{3}{2}\omega(\phi_d^r i_q^r - \phi_q^r i_d^r) + \frac{3}{2}\left(\frac{d\phi_d^r}{dt}i_d^r + \frac{d\phi_q^r}{dt}i_q^r\right), \tag{5}$$

which indicates that the output active power of SG is a result of the developed torque and of the rate of change of magnetic stored energy. Hence in the rotating reference frame, the electromagnetic torque can be expressed as:

$$T_e = \frac{3}{2}p(\phi_d{}^r i_q{}^r - \phi_q{}^r i_d{}^r),$$

(6)

where p is the number of pole pairs. According to the equivalent circuit shown above, the flux is obtained as:

$$\begin{cases} \phi_d{}^r = L_{md}{}^r i_f - (L_{ld}{}^r + L_{md}{}^r)i_d{}^r \\ \phi_q{}^r = -(L_{lq}{}^r + L_{mq}{}^r)i_q{}^r \end{cases}.$$

(7)

Based on (6) and (7) the electromagnetic torque can be rewritten as:

$$T_e = \frac{3}{2}p\left(L_{md}{}^r i_f - \Delta L_{mdq}{}^r i_d{}^r\right)i_q{}^r,$$

(8)

where $\Delta L_{mdq}{}^r = (L_{md}{}^r - L_{mq}{}^r)$. The high frequency part of T_e can thus be expressed as:

$$\hat{T}_e = \frac{3}{2}p\left(\left(L_{md}{}^r i_f - \Delta L_{mdq}{}^r \bar{i}_d{}^r\right)\hat{i}_q{}^r - \Delta L_{mdq}{}^r \hat{i}_d{}^r (\bar{i}_q{}^r + \hat{i}_q{}^r)\right),$$

(9)

where the over-line symbol represents the low frequency component of the variable while the hat symbol represents the high frequency component. The partial differentials of $\hat{T}_e$ respecting to the high frequency currents are:

$$\begin{cases} \frac{\partial \hat{T}_e}{\partial \hat{i}_d{}^r} = -\frac{3}{2}p\Delta L_{mdq}{}^r i_q{}^r \\ \frac{\partial \hat{T}_e}{\partial \hat{i}_q{}^r} = \frac{3}{2}p(L_{md}{}^r i_f - \Delta L_{mdq}{}^r i_d{}^r) \end{cases}.$$

(10)

Since in (10) the currents are the main variables, vector diagrams are drawn so that the internal current of the SG can be associated to the current that is provided by the ancillary service actuator.

Figure 2a shows the vector diagram of the SG variables under normal generative operation. Two sets of dq-frames: r-dq-frame and c-dq-frame have been drawn respectively according to the rotor position and the coupling point voltage. Therefore, in the following passage, r-d-axis, r-q-axis, c-d-axis and c-q-axis are used in short to refer to the d,q axes oriented on the rotor and on the grid voltage respectively. E_0 represents the no-load electromotive force lying on q-axis of rotor (r-q-axis). θ is the torque angle and φ is the power angle. c-dq-frame leads r-dq-frame by $(\pi/2 - \theta)$ and, θ should be an acute angle under stable condition. In case of ohmic-inductive loads, φ should be an acute angle with the armature current i lagging the terminal voltage v.

Back to the discussion of $\hat{T}_e$, as the magnetizing inductance is proportional to the reciprocal of the magnetic reluctance, the term $\Delta L_{mdq}{}^r$ in a salient pole machine is positive while in a round rotor machine it is close to zero. In normal generative operation, both $i_q{}^r$ and $(L_{md}{}^r i_f - \Delta L_{mdq}{}^r i_d{}^r)$ are positive. Referring to Equation (10), the partial differential of $\hat{T}_e$ respecting to $\hat{i}_d{}^r$ is negative for a salient pole rotor and zero for round rotor while the partial differential of $\hat{T}_e$ respecting to $\hat{i}_q{}^r$ is positive for both salient pole and round rotors. So if we are able to increase $\hat{i}_d{}^r$ and decrease $\hat{i}_q{}^r$, we can attenuate $\hat{T}_e$ as long as the armature current is located in the first quadrant of r-dq-frame, i.e., r-I. In other words, the target variation of armature current Δi should be located in quadrant r-IV when $\hat{T}_e > 0$ and in quadrant r-II when $\hat{T}_e < 0$. Without decreasing the active power of the ancillary service actuator, the reactive power can be utilized to provide the low-pass filtering of the electromagnetic torque. As the c-dq-frame is set by the voltage at the coupling point, the current which induces reactive power flow should lie on the c-q-axis. As it is shown in Figure 2, the positive part of c-q-axis locates in quadrant r-II and the negative part in quadrant r-IV. So when a load is connected to the grid, $\hat{T}_e > 0$. To attenuate $\hat{T}_e$, the ancillary service actuator injects a positive c-q-axis current Δi_{inv} to the grid and thus forces the SG to generate Δi, which is 180° shifted from Δi_{inv}, as shown in Figure 2b. Projecting Δi to r-dq-frame, we obtain:

$$\begin{cases} \Delta i_d{}^r = \Delta i \cdot cos\theta \\ \Delta i_q{}^r = -\Delta i \cdot sin\theta \end{cases} \qquad (11)$$

The change of flux can be calculated as:

$$\begin{cases} \Delta\phi_d{}^r = -(L_{ld}{}^r + L_{md}{}^r) \cdot \Delta i \cdot cos\theta \\ \Delta\phi_q{}^r = -(L_{lq}{}^r + L_{mq}{}^r) \cdot \Delta i \cdot sin\theta \end{cases}, \qquad (12)$$

indicating both fluxes in r-d-axis and r-q-axis having been weakened. The flux change depends on the amplitude and polarity of Δi, and the torque angle θ. The removed high-frequency part of the electromagnetic torque can be obtained:

$$(1 - \alpha)\hat{T}_e = \mu\frac{3}{2}p\big(\Delta L_{mdq}{}^r i \cdot \Delta i cos(2\theta + \varphi) + L_{md}{}^r i_f \Delta i sin\theta\big), \qquad (13)$$

where α is the attenuation coefficient of the electromagnetic torque; μ is the percentage of Δi in high frequency domain. The faster is the ancillary service control loop, the higher becomes the value of μ. In the case of a round rotor, it's only θ that determines the contribution of the ancillary service while in the case of salient pole rotor, the load current and the power angle matter as well. The way the ancillary service works under no-load condition is similar to changing the polarity of the controlled current Δi_{inv}.

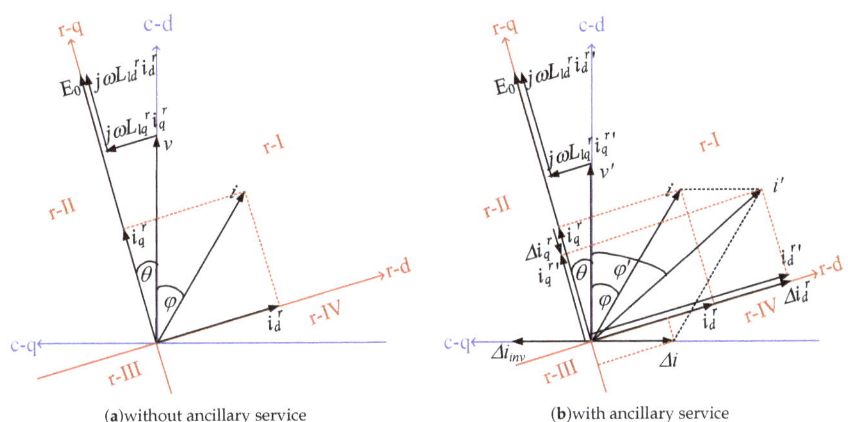

Figure 2. Vector diagrams on dq-frames set by SG and coupling point voltage.

Based on the explanations above, the frequency regulation process is summarized by the block diagram shown in Figure 3. The grid former sets and regulates the frequency of the network. Since the electromagnetic torque is a result of the excitation current and the load current, changing a part of the load current will lead to changes in the electromagnetic torque. According to Figure 2, it is possible to refer the q-axis current of the PV inverter to the same reference system of the rotor. From the comparison of the two diagrams reported it can be said that even if this q-axis component of the PV inverter current calls for only reactive power from the PV plant, it still influences the electromagnetic torque seen by the SG. In order to obtain a good result, the response speed of this control loop must be faster than that of the SG exciter. Therefore, in the PV inverter a $Q(f)$ control is implemented in order to smooth the electromagnetic torque transient without changing the active power injected in the grid. In particular, a linear relationship between reactive power and frequency deviation is implemented. It is:

$$Q = k(f^* - f) \tag{14}$$

where f^* is the reference frequency and k is tuned considering the maximum reactive power and the maximum allowed frequency variation. It is worth noting that, in a grid with distributed PV systems, the service can be performed by different devices, each one acting on the basis of its power rating. To summarize, the VSG method temporarily compensates the blanking period of the mechanical power unlike the proposed method which works on the transient of the electromagnetic torque.

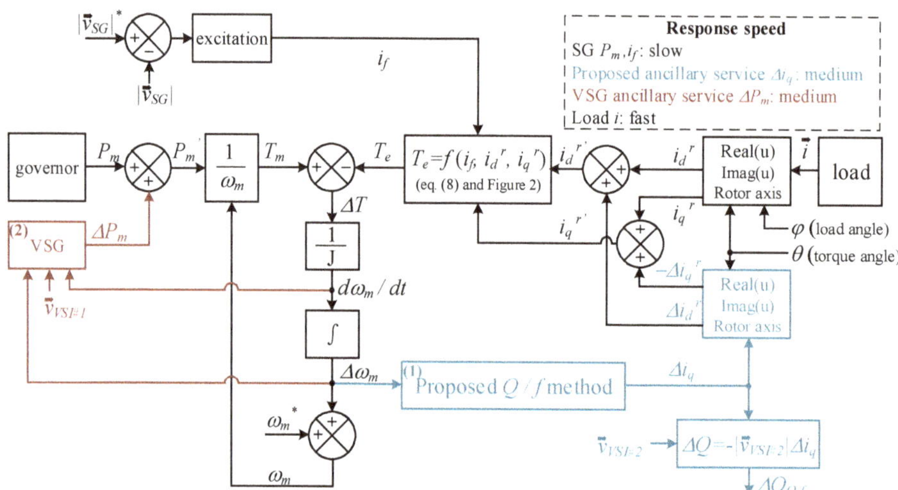

Figure 3. Frequency regulation algorithm of concerning a synchronous generator and the connected loads including both the VSG method and the proposed method.

3. Numerical Simulations: Comparison between Proposed Method and VSG

3.1. The Microgrid under Test

In this section, simulations are carried out to verify the proposed ancillary method. The system under analysis is represented in Figure 4, with power generators, loads, transmission lines and VSG equipment. The nominal frequency and line-to-line voltage of the system are respectively equal to 50 Hz and 400 Vrms. In order to have a better understanding of the analysis, most parameters are described using per unit (pu) values. Their base values are calculated and listed in Table 2.

The main power generator is a 250 kVA/400 V salient pole SG. Its detailed parameters are listed in Table 3. The SG is driven by a governor and excited by an excitation system, both of which have a much slower response compared to the power electronic devices in PV systems. The description and the parameters of the excitation and governing systems for the SG are listed in Tables 4 and 5, respectively.

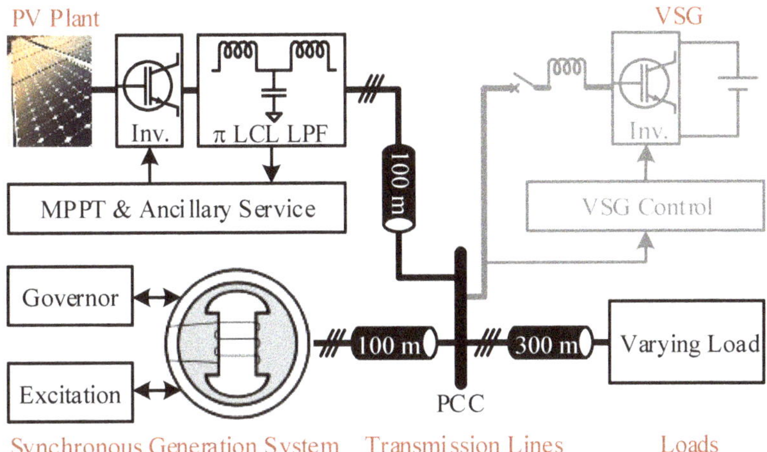

Figure 4. Block diagram of the microgrid under test.

Table 2. Base values.

Common Base Values	
Base voltage	$V_{base} = \sqrt{2/3}V_n = 326.599\,V$
Base angular frequency	$\omega_{base} = 2\pi f_n = 314.159\,rad/s$
Synchronous Generator	
Base stator current	$I_{base}^{SG} = (2S_{base}^{SG})/(3V_{base}) = 510.310\,A$
Base impedance	$Z_{base}^{SG} = V_{base}/I_{base}^{SG} = 0.640\,\Omega$
Base stator inductance	$L_{base}^{SG} = Z_{base}^{SG}/\omega_{base} = 2.037\,mH$
PV Plant	
Base output current	$I_{base}^{PV} = (2S_{base}^{PV})/(3V_{base}) = 204.124\,A$
Base impedance	$Z_{base}^{PV} = V_{base}/I_{base}^{PV} = 1.600\,\Omega$
Base inductance	$L_{base}^{PV} = Z_{base}^{PV}/\omega_{base} = 5.093\,mH$
Base capacitance	$C_{base}^{PV} = 1/(\omega_{base}Z_{base}^{PV}) = 1.989\,mF$
VSG	
Base output current	$I_{base}^{VSG} = (2S_{base}^{VSG})/(3V_{base}) = 51.031\,A$
Base impedance	$Z_{base}^{VSG} = V_{base}/I_{base}^{VSG} = 6.4\,\Omega$
Base inductance	$L_{base}^{VSG} = Z_{base}^{VSG}/\omega_{base} = 20.372\,mH$
Base capacitance	$C_{base}^{VSG} = 1/(\omega_{base}Z_{base}^{VSG}) = 0.497\,mF$

Table 3. Parameters of the SG in the microgrid.

Configuration			
Rotor type	Salient-pole	Stator windings	wye connection
Pole pairs	2	Friction factor	0.16 Nms
Moment of inertia	3.553 kgm^2		
Nominal Ratings			
Power	250 kVA	Voltage(rms line-line)	400 V
Frequency	50 Hz	N_s/N_f	0.1
Parameters (pu) (base values are calculated in Table 2–Synchronous Generator)			
Stator resistance	0.026	Stator leakage inductance	0.090
Magnetizing inductance (d)	2.750	Magnetizing inductance (q)	2.350
Field winding resistance	0.094	Field winding leakage inductance	147.262
Damping resistance (d)	0.292	Damping inductance (d)	1.982
Damping resistance (q)	0.066	Damping inductance (q)	0.305

Table 4. Parameters of the excitation system for the SG.

Approximate model	PI regulator
Setpoint	Amplitude of line-line voltage = 566 (V)
Output Signal	Field voltage (V)
Proportional coefficient	15
Integral coefficient	80
Output range	[0, 240] (V)

Table 5. Parameters of the governing system for the SG.

Approximate model	PI regulator
Setpoint	Rotor angular frequency = 157 (rad/s)
Output Signal	Mechanical torque (Nm)
Proportional coefficient	70
Integral coefficient	51
Output range	[0, 1600] (Nm)

The second power generator unit is a PV plant. The nominal active power at standard test condition (STC) is 100 kW. The detailed parameters of the PV plant are listed in Table 6.

Table 6. Parameters of the PV plant in the microgrid.

Nominal Ratings of the PV plant			
Power	100 kW	Voltage(rms line-line)	400 V
Frequency	50 Hz	Switching frequency	20 kHz
Parameters of the PV array @ STC *			
Maximum Power	100 kW		
Voltage @ MPP	273.5 V	Current @ MPP	368.3 A
Parameters of the output LCL Filter (pu) (base values are calculated in Table 2–PV Plant)			
Inverter side inductor	15.669×10^{-3}	Inverter side resistor	6.250×10^{-6}
Grid side inductor	9.424×10^{-3}	Grid side resistor	6.250×10^{-6}
Shunt capacitor	0.272	Shunt resistor	0.049

* Standard test condition: 1000 W/m^2 and 25 °C.

The VSG is realized by a three-phase inverter whose nominal power is 25 kW. The detailed parameters are listed in Table 7.

Table 7. Parameters of the VSG in the microgrid.

Nominal Ratings of the VSG			
Power	25 kW	Voltage(rms line-line)	400 V
Frequency	50 Hz	DC side voltage	1 kV
Parameters of the output LCL Filter (pu) (base values are calculated in Table 2–VSG)			
Inverter side inductor	3.917×10^{-3}	Inverter side resistor	1.563×10^{-6}
Grid side inductor	2.356×10^{-3}	Grid side resistor	1.563×10^{-6}
Shunt capacitor	1.089	Shunt resistor	0.012

Finally, the parameters of the transmission lines are described in Table 8.

Table 8. Parameters of the transmission lines.

R/km	L/km	l_{SG-PCC}	$l_{Load-PCC}$	l_{PV-PCC}
78 mΩ/km	238 µH/km	0.1 km	0.3 km	0.1 km

3.2. Control Methods

As previously discussed, the grid-tied inverter of the PV plant can be controlled to implement the proposed ancillary service as illustrated in Figure 3. The PV system is controlled to inject in the grid the active power given by the MPPT algorithm while the reactive power reference, Q, is set to zero when the frequency is at rated value. In the test of the proposed ancillary service, the reactive power is regulated by means of the current on c-q-axis related to the frequency transient rate. The VSG method is, instead, executed by adding a virtual inertia and friction factor to the active power control loop to emulate the mechanical behavior of the rotor.

3.2.1. Proposed Q/f Control

The control algorithm of grid-tied inverter of the PV plant is proposed as shown in Figure 5. It consists of active power control loop and reactive power control loop. In the active power control, perturb & observe MPPT is firstly performed to determine the right duty cycle of the boost converter. Then the output voltage of the boost converter is stabilized by a feedback control loop using a PI regulator with $K_p = 7$ and $K_i = 800$. The set point of this loop is 500 V and the output will be the set point of c-d-axis current in pu values limited between $[-1.5, 1.5]$ pu. The reactive power control aims at stabilizing frequency. When frequency deviates from the set point 50 Hz, a c-q-axis current (in pu values) is set as a c-q-axis current reference by a positive coefficient $D_{Q/f} = 0.5$:

$$\begin{cases} i_q{}^*(k) = D_{Q/f}\big(f^* - f(k)\big) & v \in [360\text{V}, 440\text{V}] \\ i_q{}^*(k) = i_q{}^*(k-1) & v \notin [360\text{V}, 440\text{V}] \end{cases} . \tag{15}$$

The set point of c-q-axis is kept between limits $[-1.5, 1.5]$ pu. With the two set points of currents, the current loop is controlled by a PI regulator with $K_p = 0.3$ and $K_i = 20$. The output signals will be considered as the set points of c-d-axis and c-q-axis voltages in pu values and be limited between $[-2, 2]$ pu. Feed-forward control concerning the output filter is added to speed up the control loop.

The regulators have been tuned by means of a trial and error procedure in order to obtain a fast and stable response by each control loop. The choice of the droop coefficient $D_{Q/f}$ is done to have a pu quadrature current 0.5 when a 1Hz (i.e., 2%) deviation occurs on the grid frequency.

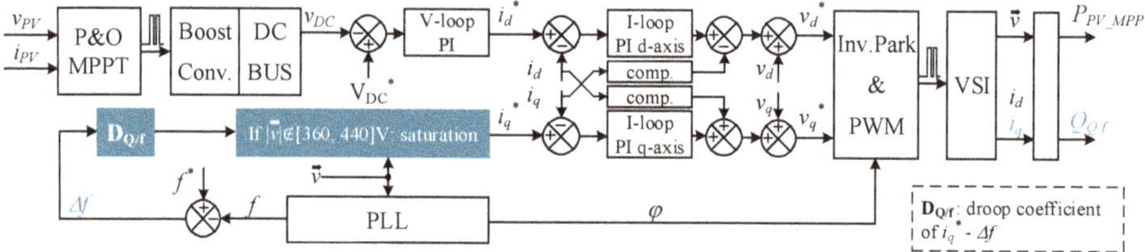

Figure 5. Control block diagram of the PV system.

3.2.2. VSG Control

At the same time, a parallel comparison is made between the proposed method and VSG. The VSG is implemented by inserting an extra voltage source inverter (VSI) supplied by a DC voltage source. The control algorithm is depicted by the block diagram shown in Figure 6. Similar to the swing equation of the SG, the virtual inertia and the virtual friction factor amplify the ROCOF and the frequency deviation, respectively. A corresponding electric power is thus generated to emulate the change of mechanical power of the SG in order to dampen the frequency transient. Here in the test, a virtual inertia J = 15 kgm^2 and a friction factor F = 33 Nms are used to replicate the change of the mechanical torque. Therefore, the set point is the nominal mechanical angular frequency and the output is the set point of mechanical torque (both in SI values). Based on the mechanical angular frequency and the base values of the VSG shown in Table 2, the set point of the active power in pu value can be obtained which will be limited between $[-1, 1]$ pu. Since the main focus is on frequency, the set point of the reactive power of VSG is set equal to zero. With set points of active and reactive power and the measured voltage, set points of c-d-axis and c-q-axis currents can be calculated. Feedback loops are controlled by PI regulators with $K_p = 0.3$ and $K_i = 20$. The outputs are set to the set points of c-d-axis and c-q-axis voltages in pu values limited between $[-2, 2]$ pu. Feed-forward control is applied to compensate the voltage drop across the output filter. The regulators have been tuned by means of a trial and error procedure in order to obtain a fast and stable response by each control loop.

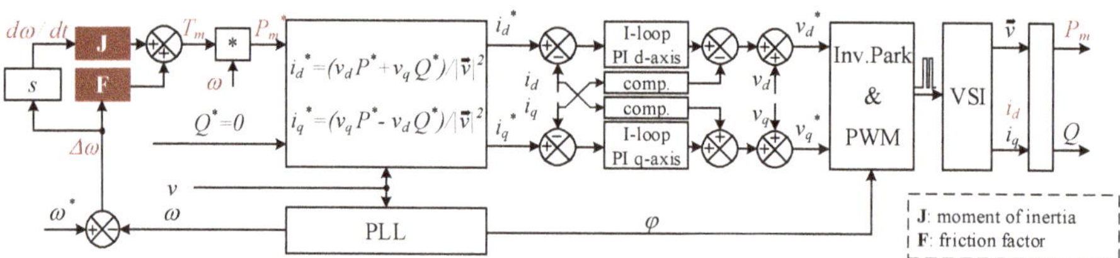

Figure 6. Control block diagram of the VSG.

3.2.3. Test Condition and Results

In Sections 3.2.4 and 3.2.2, the droop coefficient of the Q/f control is set to 0.5 pu/Hz and the virtual inertia and the friction factor are set to 15 kgm^2 and 33 Nms for VSG control. Under the condition of such parameters, the two ancillary services can achieve the same attenuation of the frequency overshooting during a transient. In this way, the two methods can be compared from the working principle aspect.

The simulation test starts from a steady state where load power is 175.8 kW–6.1 kvar. Referring to the base values of the SG, the load power is 0.704–0.024 j pu. The PV plant is generating 100 kW which means 1 pu referring to PV plant. Both ancillary services are not activated. In order to produce a frequency transient, an extra load of 27 kW–1.4 kvar (0.108–0.006 j pu referring to the SG) is connected to the micro grid at 0 s and disconnected at 10th s. The connection and disconnection is done by a circuit breaker. Therefore, under-frequency and over-frequency situations are provided.

The test is repeated three times. At the first time, neither of the ancillary services is activated. In the following plots, this case is named as No AS and it is taken as the reference. At the second time, just the proposed ancillary service, i.e., the Q/f control is activated. This case is named as Proposed AS. Finally, the VSG ancillary service is activated. This case is named as VSG AS.

The frequency transients are shown in Figure 7a. Even though the maximum deviations of frequency are comparable for VSG and the proposed ancillary service, the frequency transients are still different. According to the test results, both ancillary services are effective on reducing frequency over-shooting. However, the two methods are distinguished from each other by the transient curves.

Figure 7b describes the power provided by the two ancillary services. In Q/f control, it is the reactive power that is responsible for the frequency transient mitigation where in VSG control, the active power is in charge. As shown in Figure 7b a 10% of additional reactive power is required to the inverter of the RES. In order to be able to exchange this reactive power also when the active power is maximum, the inverter has to be oversized. Anyway, an additional 10% of reactive power implies only a 0.5% increasing of the apparent power. Therefore, the additional cost to oversize the inverter (by 0.5%) can be considered negligible and it is possible to state that the service can be obtained with almost null costs. In the VSG test, ΔP performs the compensation process with the slow response of the governer. ΔQ in Q/f method controls the flux inside SG and thus smoothing the voltage recovery, which is shown in Figure 7c. Summarizing, the proposed ancillary service performances are comparable to those of a VSG in terms of limitation of minimum and maximum frequencies during the transients. Nevertheless, the recovery time of both frequency and voltage is slowed by the proposed ancillary service. Even if this seems a disadvantage, it is worth highlighting that this is obtained without needing any energy reserve and this makes the proposed service implementable in all the RES devices distributed in the grid. This is the main advantage of the proposed algorithm in comparison with the traditional VSG.

Figure 8 shows the transients of the internal variables of the SG based on the c-dq-frame. The armature currents are regulated by the ancillary services. Relating to the reference current obtained in the reference test, the changing trends of i_d and i_q being regulated by Q/f method and VSG method are different. Therefore, the resultant electromagnetic torques T_e of the two ancillary services have different shapes. However, both torques are smoother than that of the reference test, giving more time to the governor system to follow the change of load.

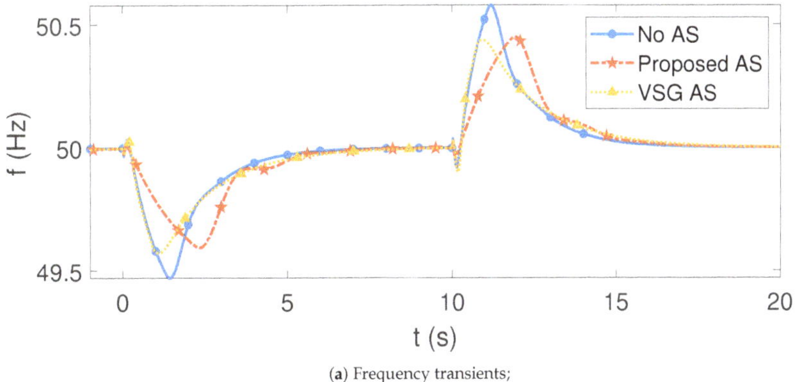

(a) Frequency transients;

Figure 7. *Cont.*

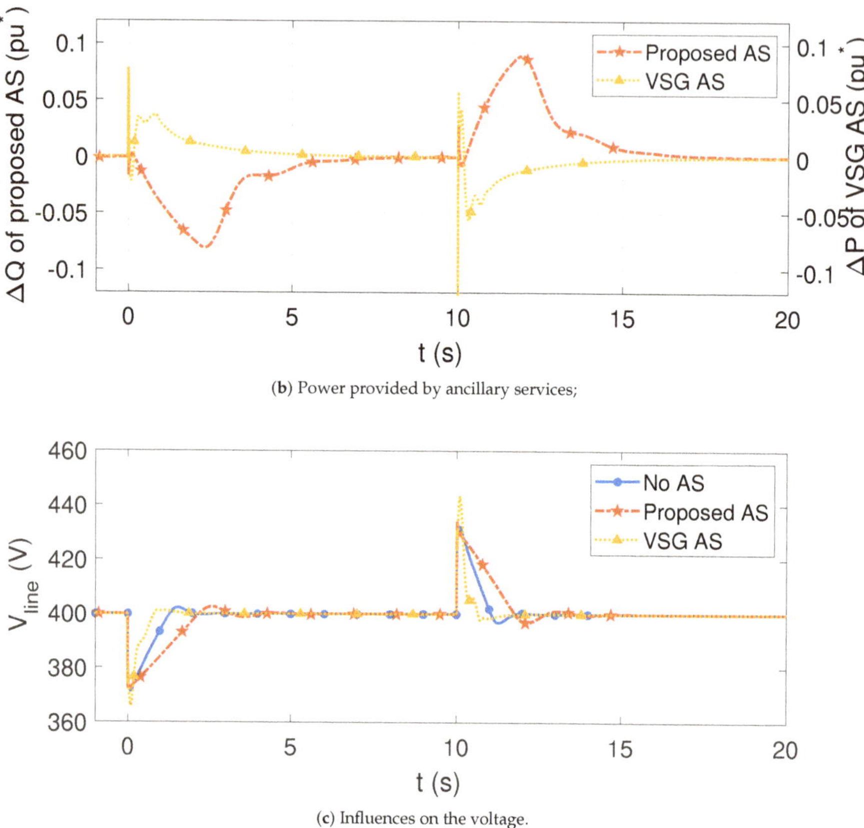

(**b**) Power provided by ancillary services;

(**c**) Influences on the voltage.

Figure 7. Offline simulation test results of the proposed ancillary service (AS) in comparison with VSG.
* pu values are obtained according to the base values of the synchronous generation system in Table 2.

With quite close performances of alleviating frequency deviation, the proposed method is shown to be more efficient owing to the sole use of reactive power. In other words, the proposed method does not ask for an extra reserve to provide the requested active power. From the budget and simplicity point of view, the proposed ancillary service is a viable choice for the existing networks.

The main advantage of the method is that it can be implemented on every grid-connected inverter and works without affecting the functionalities of the MPPT. Requiring no additional power reserves, this methodology reduces costs of installation and can be flexibly integrated into the existing equipment.

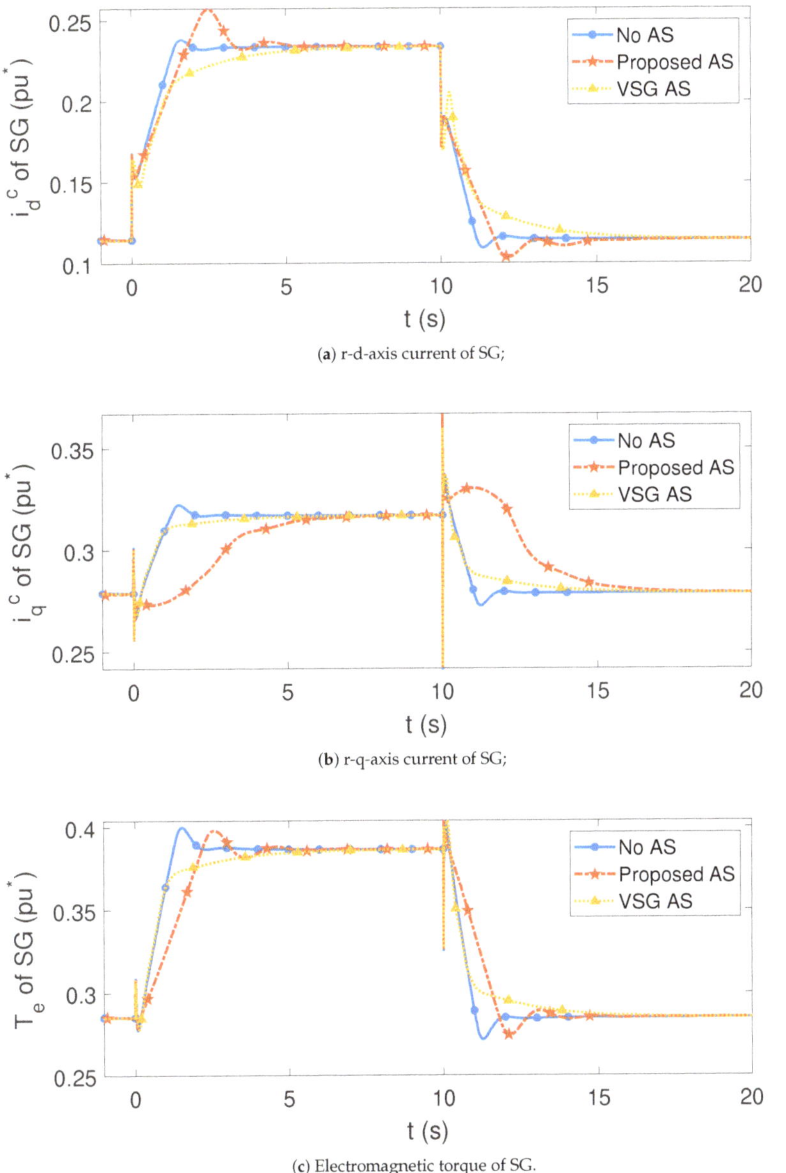

(**a**) r-d-axis current of SG;

(**b**) r-q-axis current of SG;

(**c**) Electromagnetic torque of SG.

Figure 8. Internal changes of the SG due to the proposed ancillary service. * pu values are obtained according to the base values of the synchronous generation system in Table 2.

3.2.4. Stability Analisys of the Proposed Q/f Control

In order to test the local stability of the proposed Q/f control we chose to use the indirect Lyapunov method for nonlinear systems. This method consists in linearizing the nonlinear system around an equilibrium point and assess its local stability for small perturbations. In our case, we want to assess the mechanical frequency stability of the SG at 50 Hz. The linearization was performed using the linear analysis tool of Matlab/Simulink software. Moreover, for the stability analysis all the saturations of the regulators were removed and the VSG is not connected. In the simulink model, we had to select one input perturbation point and one output measurement point in order to obtain the linearized closed loop transfer function between the mechanical frequency of the SG and the reference one. In fact, a perturbation in the reference frequency acts on both the governor of the SG and the proposed Q/f control. The poles placement of this closed loop transfer function is dependent on several parameters, among which, the value of the droop coefficient $D_{Q/f}$ that we want to assess for the stability analysis. Therefore, this was varied between 0 (proposed control not active) and 15 pu/Hz with a step of 0.5 pu/Hz. Since, the system was simulated using a discrete solver the poles are in the z-domain. As is well known, a nonlinear time invariant discrete system, trimmed at an equilibrium point and for small perturbations, is stable if and only if all the poles of the linearized system have an amplitude less than one, i.e., they are into the circumference of unitary radius. Figure 9 shows the zero-pole map of the closed loop transfer function for the different droop coefficient values. We can note that the region of the map in which some poles are out of the circumference is near to 1. Figure 10 show a zoom of such region. From this figure, we can see that for increasing values of the droop coefficient the poles are moving towards the boundary of the circumference up to pass it for values higher than 11 pu/Hz. This means that for droop coefficients greater than 11 pu/Hz the system becomes unstable; for droop coefficients less than 11 pu/Hz the system is locally stable, i.e., only for small perturbations. In order to assess the convergence domain, i.e., for which values of perturbation the system is stable, we should use other stability methods that for our system can be very difficult to apply. On the other hand, the transfer function for the chosen value of the droop coefficient (0.5 pu/Hz) has the poles far enough from the boundary of the circumference. Moreover, the actual system contains several saturations in the controllers helping in stabilizing the system response. Therefore, it is possible to state that the proposed service is stable if the droop coefficient is chosen much lower than the stability limit. In the paper a value 20 times lower than the limit was used obtaining a stable answer from the system.

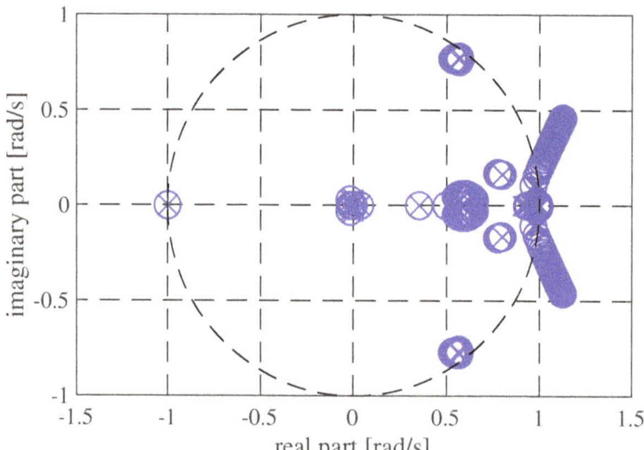

Figure 9. Zero-pole map. Poles (crosses); zeros (circles).

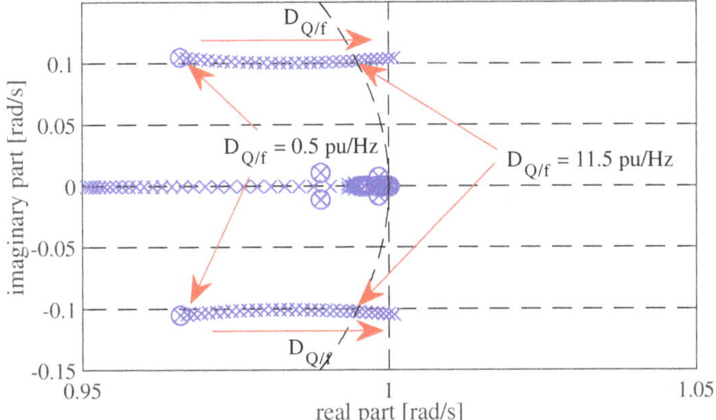

Figure 10. Zoom of the zero-pole map. Poles (crosses); zeros (circles).

4. CHIL Simulation Results

4.1. CHIL Platform of Microgrid

In order to test the effectiveness of the proposed ancillary service on real processor, a CHIL platform is built on the basis of a real-time controller dSPACE [38] and a real-time simulator Typhoon HIL [39]. Figure 11 illustrates the general construction of the micro grid. The complete PV control is implemented by the real embedded system while the rest of the system including the model and other control units are simulated in real-time by the simulator.

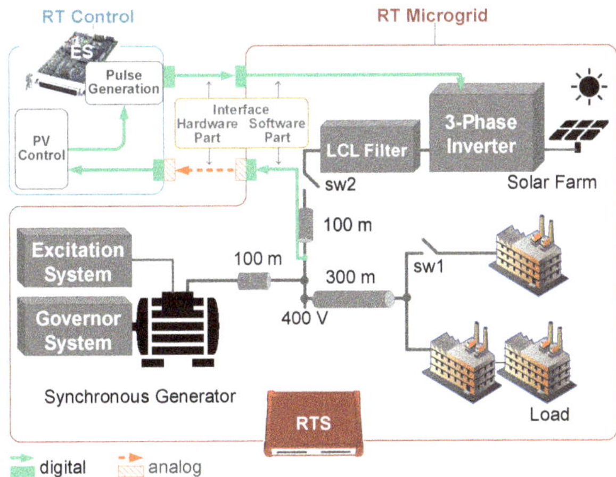

Figure 11. CHIL simulation structure of the microgrid.

4.1.1. Real-Time Microgrid

To reduce the total computation burden and to make a better usage of the hardware resources, the model is subdivided into two parts considering the PV system in one core and the rest in another using an ideal transformer model (ITM) as the interface algorithm [40]. The ITM is placed in the LCL output filter of the PV inverter (Figure 12). The single-line diagram then is used to clarify the description. A voltage amplified ITM is placed within the PV inverter at the primary side and the grid at the secondary side. The stability of the system is affected by the value of the impedances on both sides of ITM as shown in [41]. Therefore, more attention needs to be paid while setting the ITM parameters. After circuit partition, the computation burden is greatly reduced from 145 % by one core to 25 % and 3 % respectively by two cores.

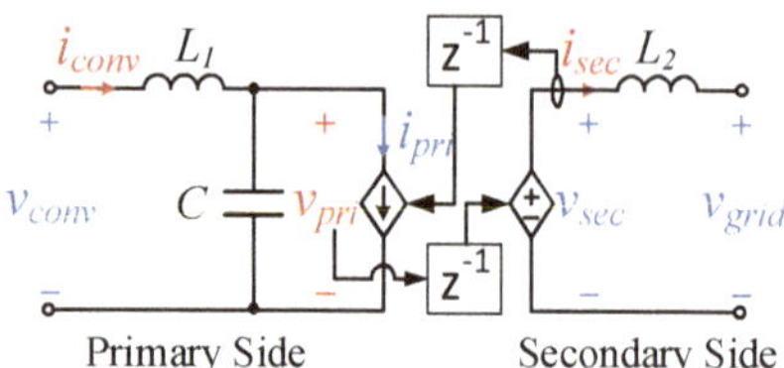

Figure 12. Equivalent circuit of the partitioned LCL filter used in the Real-time simulation.

The fundamental time step is defined as 1 µs, which is the reference clock to synchronize all parts of the system as well. In the micro grid, the SG is the only source of natural inertia; furthermore, since the proposed approach handles the reactive power flow from the PV plant to mitigate the frequency transients, the system's response has to be faster than the excitation system of the SG. The electrical part of the machine is modeled by a fifth-order state-space model in a synchronously rotating d-q coordinates; the mechanical part is modeled by a second-order state-space model [42]. The execution rate of the simulation of the synchronous generation system is defined as 5 kHz and the parameters are listed in Tables 3–5.

The PV plant is composed of DC side voltage source (PV panels), 3-phase inverter and LCL filter. The state-space variables are calculated at every fundamental step. The inverter gate driving signals are calculated at 1 MHz and therefore, they are generated with an oversampling frequency of 50 MHz, guaranteeing high fidelity of the pulsed signals even under fast switching and narrow duty cycle conditions. The features of the PV plant are listed in Table 6.

The loads and transmission lines are simulated at 1 MHz. The loads consist of a permanent 175.8 kW–6.1 kvar load and an optional 27 kW–1.4 kvar load which creates the frequency transients by connection and disconnection actions. For transmission the equivalent three phase RL modeled overhead lines are used. The related parameters are shown in Table 8. The synchronous generation system, PV system and the loads are joined at the PCC via transmission lines of 100 m, 100 m and 300 m respectively.

4.1.2. Interface and Real-Time Control

As previously mentioned, the PV control algorithm is executed by an embedded system in real-time out of the grid simulator, so an interface is required to join these two devices both physically and logically. The inputs of the control are voltages and currents measured at the PCC and the outputs of the control are the driving signals for the PV inverter. In real implementation, the voltages and currents are firstly measured and transformed by transducers. Before being sent to the controller, these analog signals are normally amplified or attenuated and filtered by conditioning circuits and finally converted into digital signals. On the other hand, the generated gate driving signals are sent to the driving circuit of the inverter where the non-ideal switching of the semiconductors takes place. Aiming at replicating the impact of the sampling devices, a software interface is created inside the real-time simulator, as it is illustrated in Figure 13. The PCC variables are firstly sampled by the maximum rate available in the model, i.e., 1 MHz to avoid aliasing issues. Then the samples go through a set of 1st-order low pass filters (LPFs) and absolute time delays which are executed at 50 kHz, to represent the limited bandwidth, anti-aliasing handling and response time of the measurement system. After the functional part, the physical connection is achieved by the analog-to-digital converters (ADCs), digital-to-analog converters (DACs) and digital inputs/outputs (DI/Os) of the two real-time devices. The conditioned PCC variables are amplified by DACs at 1 MHz. Compared with the receiver ADCs at controller side (10 kHz), the PCC variables are quasi-continuous. The modulation wave generated by the control algorithm at 10 kHz is then transformed into pulses by a slave dsp whose resolution is 100 ns. And finally the pulses are over-sampled by 50 MHz closely tracking the expected duty cycle.

The control algorithm follows the block diagram shown in Figure 5. The switching frequency is set at 20 kHz. With 100 ns resolution of the carrier signal, the output's duty cycle will have a resolution of 0.2 %. The reading commands of ADCs and the updating of the modulation wave are arranged at the beginning of the code. To a large extent, this fixes the time baseline when the variables are taken and sent within a control period, which avoids the unexpected high frequency harmonics caused by the embedded system, but introduces one-step delay to the actual control at the same time.

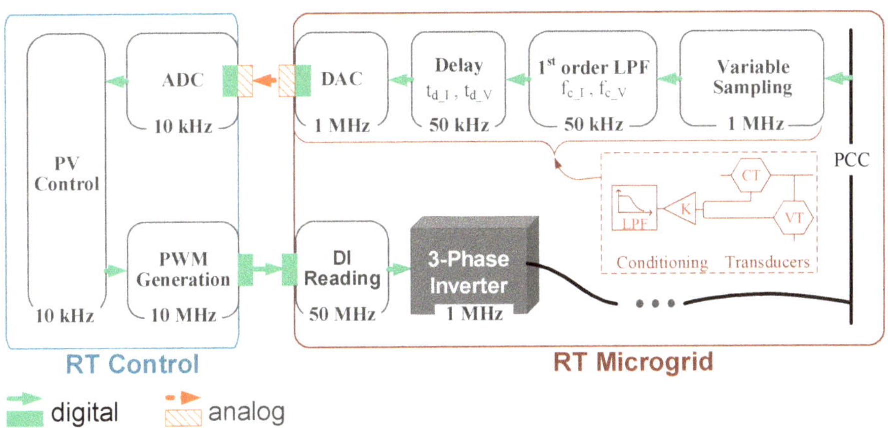

Figure 13. CHIL interfacing method.

4.2. Exposed Control Problem

Taking a brief look at the CHIL test results reveals a power offset (shown in Figure 14), posing unexpected power flow in the grid. This roots in the fact that theoretical ancillary service was proposed under ideal conditions where the variables had infinite bandwidth, unit gain and zero latency. However, this is not valid in CHIL and in practice. As it is explained in the previous subsection, there are measurement devices between PCC and control system featuring limited bandwidth, delay and filtering effect. The PLL introduces phase delay and time delay as well. These weak points influence the precision of the phase angles based on which the voltages and currents are transformed from rotating values to static ones. The consequence inspires control system designers to consider the effect of the measurement chain in practical control design. In this case, the shifted phase angle caused by the non-ideal signal transmission procedure and the PLL can be corrected by means of equivalent delay compensation. However, it is worth noting that the signal transmission chain is a mix of absolute time delays and frequency-dependent phase angle lags. Frequency dependent lags are estimated at rated frequency, i.e., 50 Hz. After integrating the measurement delay into the control algorithm, the reactive power seen by the controller is almost identical to the measured value in the micro grid, as plotted in Figure 15.

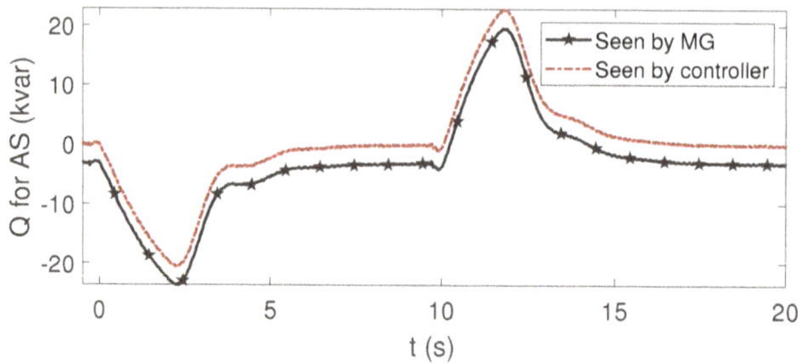

Figure 14. CHIL PV; reactive power seen by the controller (maroon) and seen by the microgrid (black).

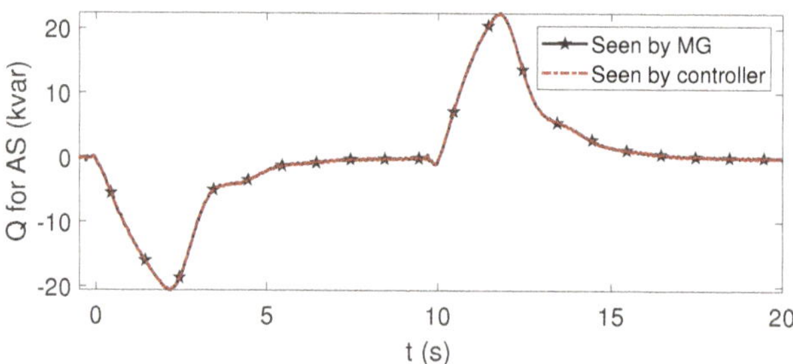

Figure 15. CHIL PV; reactive power seen by the controller(maroon) and seen by the micro grid (black) after compensation of the measurement chain.

4.3. Final Test Results

The simulation initiates from the steady state where the 175.8 kW–6.1 kvar load is connected to the PCC. The transients to be observed are induced by the connection and disconnection of the 27 kW–1.4 kvar load at 0 s and 10th s respectively. Figure 16 reports the CHIL test results in comparison with the offline simulation. The reference test case does not include the activation of any ancillary service. The results are marked as No AS CHIL and plotted in blue. The CHIL test results of the proposed Q/f ancillary service are marked as Proposed AS CHIL and plotted in yellow. The offline simulation test results are shown as well in order to compare the test between the offline simulation and CHIL. It is marked as Proposed AS Sim and plotted in red.

The CHIL test results prove that the proposed method is able to attenuate the frequency over-shooting caused by either the load connection or disconnection. In particular, the frequency undershoot is reduced by 0.15 Hz corresponding to an improvement of 27.3% considering that in the base case the undershoot is 0.55 Hz. Moreover, the overshoot is reduced by 0.14 Hz corresponding to 23.3% of the base case variation equal to 0.6 Hz. The ancillary service has a negative effect on the voltage transient as it was expected by the theoretical analysis: the voltage recovery process is prolonged, but the over-shooting peak is not significantly increased.

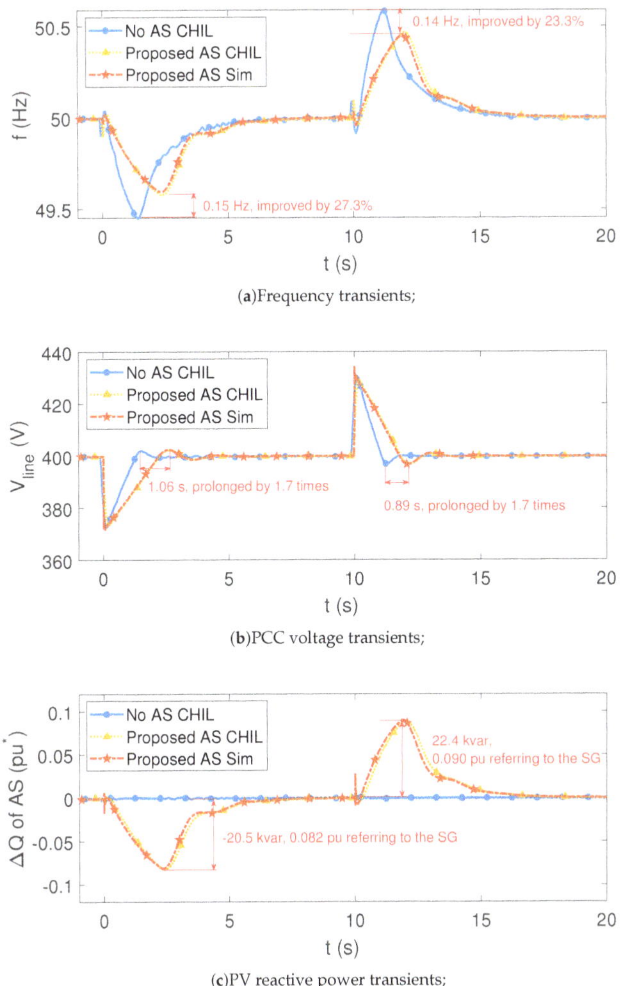

(**a**)Frequency transients;

(**b**)PCC voltage transients;

(**c**)PV reactive power transients;

Figure 16. Dynamic performances of the systems with (yellow) and without (blue) ancillary service in CHIL versus the ancillary service test results in offline simulation (red). * pu values are obtained according to the base values of the synchronous generation system in Table 2.

5. Conclusions

This paper proposes a frequency-assisting ancillary service. It works in the context of modern micro grid with reduced inertia and it can be implemented in the distributed RESs. The proposed algorithm presents performances comparable to those of a traditional VSG in mitigating frequency transients due to load variations. Nevertheless, contrarily to VSG, the proposed service does not require energy reserve to be implemented since it does not affect the active power exchanged by the RES inverter with the grid. This represents an important added value of the proposed algorithm since it is implementable in all the RES devices distributed in the grid with almost null additional costs.

The ancillary service is explained theoretically by equations and verified in simulation. The performance is compared with that of the VSG method. Even though both of the ancillary services are able to achieve improvement on frequency transients, the working principles behind them are quite distinct and thereby, the requested types of power used for the ancillary services differ as well. Finally, a CHIL micro grid system is constructed to test the algorithm in an embedded system. However, it was noticed that without cautious consideration related to the non-ideal measurement and the delay caused by data processing, the control algorithm heads toward the performance deterioration and unexpected power loss. After modification, the ancillary service on processor proves to be effective on damping the over-shooting of frequency.

Author Contributions: Conceptualization, Y.H., S.B., L.P. and G.G.; methodology, Y.H., S.B., L.P. and G.G.; software, Y.H. and S.B.; validation, L.P. and G.G.; formal analysis, Y.H., S.B., L.P. and G.G.; investigation, Y.H. and S.B.; resources, L.P. and G.G.; data curation, Y.H. and S.B.; writing–original draft preparation, Y.H., S.B.; writing–review and editing, Y.H., S.B., L.P. and G.G.; supervision, L.P. and G.G.; All authors have read and agreed to the published version of the manuscript.

Funding: This research received no external funding.

Conflicts of Interest: The authors declare no conflict of interest.

References

1. Chicco, G.; Mancarella, P. Distributed multi-generation: A comprehensive view. *Renew. Sustain. Energy Rev.* **2009**, *13*, 535–551. [CrossRef]

2. Llaria, A.; Curea, O.; Jiménez, J.; Camblong, H. Survey on microgrids: Unplanned islanding and related inverter control techniques. *Renew. Energy* **2011**, *36*, 2052–2061. [CrossRef]

3. Georgilakis, P.S.; Hatziargyriou, N.D. A review of power distribution planning in the modern power systems era: Models, methods and future research. *Electr. Power Syst. Res.* **2015**, *121*, 89–100. [CrossRef]

4. D'Arco, S.; Suul, J.A.; Fosso, O.B. A Virtual Synchronous Machine implementation for distributed control of power converters in SmartGrids. *Electr. Power Syst. Res.* **2015**, *122*, 180–197. [CrossRef]

5. Lopes, J.P.; Hatziargyriou, N.; Mutale, J.; Djapic, P.; Jenkins, N. Integrating distributed generation into electric power systems: A review of drivers, challenges and opportunities. *Electr. Power Syst. Res.* **2007**, *77*, 1189–1203. Distributed Generation. [CrossRef]

6. Carrasco, J.M.; Franquelo, L.G.; Bialasiewicz, J.T.; Galvan, E.; PortilloGuisado, R.C.; Prats, M.A.M.; Leon, J.I.; Moreno-Alfonso, N. Power-Electronic Systems for the Grid Integration of Renewable Energy Sources: A Survey. *IEEE Trans. Ind. Electron.* **2006**, *53*, 1002–1016. [CrossRef]

7. Blaabjerg, F.; Teodorescu, R.; Liserre, M.; Timbus, A.V. Overview of Control and Grid Synchronization for Distributed Power Generation Systems. *IEEE Trans. Ind. Electron.* **2006**, *53*, 1398–1409. [CrossRef]

8. Rocabert, J.; Luna, A.; Blaabjerg, F.; Rodríguez, P. Control of Power Converters in AC Microgrids. *IEEE Trans. Power Electron.* **2012**, *27*, 4734–4749. [CrossRef]

9. Hossain, M.A.; Pota, H.R.; Hossain, M.J.; Blaabjerg, F. Evolution of microgrids with converter-interfaced generations: Challenges and opportunities. *Int. J. Electr. Power Energy Syst.* **2019**, *109*, 160–186. [CrossRef]

10. Yang, Y.; Blaabjerg, F.; Wang, H. Low-Voltage Ride-Through of Single-Phase Transformerless Photovoltaic Inverters. *IEEE Trans. Ind. Appl.* **2014**, *50*, 1942–1952. [CrossRef]

11. Stetz, T.; Marten, F.; Braun, M. Improved Low Voltage Grid-Integration of Photovoltaic Systems in Germany. *IEEE Trans. Sustain. Energy* **2013**, *4*, 534–542. [CrossRef]

12. Hossain, M.; Pota, H.; Hossain, M.; Haruni, A. Active power management in a low-voltage islanded microgrid. *Int. J. Electr. Power Energy Syst.* **2018**, *98*, 36–47. [CrossRef]

13. Engler, A.; Soultanis, N. Droop control in LV-grids. In Proceedings of the 2005 International Conference on Future Power Systems, Amsterdam, NL, USA, 18 November 2005; p. 6. [CrossRef]

14. Yu, X.; Khambadkone, A.M.; Wang, H.; Terence, S.T.S. Control of Parallel-Connected Power Converters for Low-Voltage Microgrid—Part I: A Hybrid Control Architecture. *IEEE Trans. Power Electron.* **2010**, *25*, 2962–2970. [CrossRef]

15. Moradi, M.H.; Eskandari, M.; Hosseinian, S.M. Cooperative control strategy of energy storage systems and micro sources for stabilizing microgrids in different operation modes. *Int. J. Electr. Power Energy Syst.* **2016**, *78*, 390–400. [CrossRef]

16. Vandoorn, T.L.; Kooning, J.D.D.; Meersman, B.; Zwaenepoel, B. Control of storage elements in an islanded microgrid with voltage-based control of DG units and loads. *Int. J. Electr. Power Energy Syst.* **2015**, *64*, 996–1006. [CrossRef]

17. De Brabandere, K.; Bolsens, B.; Van den Keybus, J.; Woyte, A.; Driesen, J.; Belmans, R. A Voltage and Frequency Droop Control Method for Parallel Inverters. *IEEE Trans. Power Electron.* **2007**, *22*, 1107–1115. [CrossRef]

18. Guerrero, J.M.; De Vicuna, L.G.; Matas, J.; Castilla, M.; Miret, J. Output impedance design of parallel-connected UPS inverters with wireless load-sharing control. *IEEE Trans. Ind. Electron.* **2005**, *52*, 1126–1135. [CrossRef]

19. He, J.; Li, Y.W. Analysis, Design, and Implementation of Virtual Impedance for Power Electronics Interfaced Distributed Generation. *IEEE Trans. Ind. Appl.* **2011**, *47*, 2525–2538. [CrossRef]

20. Kim, J.; Guerrero, J.M.; Rodriguez, P.; Teodorescu, R.; Nam, K. Mode Adaptive Droop Control With Virtual Output Impedances for an Inverter-Based Flexible AC Microgrid. *IEEE Trans. Power Electron.* **2011**, *26*, 689–701. [CrossRef]

21. Anderson, P.M.; Fouad, A.A. *Power System Control and Stability*; IEEE Press: Piscataway, NJ, US, 2003.

22. Pulendran, S.; Tate, J. Energy storage system control for prevention of transient under-frequency load shedding. In Proceedings of the 2017 IEEE Power Energy Society General Meeting, Chicago, IL, USA, 16–20 July 2017; p. 1. [CrossRef]

23. Wen, Y.; Li, W.; Huang, G.; Liu, X. Frequency dynamics constrained unit commitment with battery energy storage. In Proceedings of the 2017 IEEE Power Energy Society General Meeting, Chicago, IL, USA, 16–20 July 2017; p. 1. [CrossRef]

24. Barcellona, S.; Huo, Y.; Niu, R.; Piegari, L.; Ragaini, E. Control strategy of virtual synchronous generator based on virtual impedance and band-pass damping. In Proceedings of the 2016 International Symposium on Power Electronics, Electrical Drives, Automation and Motion (SPEEDAM), Anacapri, Italy, 22–24 June 2016; pp. 1354–1362. [CrossRef]

25. Liu, J.; Hossain, M.; Lu, J.; Rafi, F.; Li, H. A hybrid AC/DC microgrid control system based on a virtual synchronous generator for smooth transient performances. *Electr. Power Syst. Res.* **2018**, *162*, 169–182. [CrossRef]

26. Tan, J.; Zhang, Y. Coordinated Control Strategy of a Battery Energy Storage System to Support a Wind Power Plant Providing Multi-Timescale Frequency Ancillary Services. *IEEE Trans. Sustain. Energy* **2017**, *8*, 1140–1153. [CrossRef]

27. Johnson, J.; Neely, J.C.; Delhotal, J.J.; Lave, M. Photovoltaic Frequency–Watt Curve Design for Frequency Regulation and Fast Contingency Reserves. *IEEE J. Photovolt.* **2016**, *6*, 1611–1618. [CrossRef]

28. Ochoa, D.; Martinez, S. Fast-Frequency Response Provided by DFIG-Wind Turbines and its Impact on the Grid. *IEEE Trans. Power Syst.* **2017**, *32*, 4002–4011. [CrossRef]

29. Schleif, F.R.; Hunkins, H.D.; Martin, G.E.; Hattan, E.E. Excitation Control to Improve Powerline Stability. *IEEE Trans. Power Appar. Syst.* **1968**, *PAS-87*, 1426–1434. [CrossRef]

30. Larsen, E.V.; Chow, J.S. *SVC Control Design Concepts for Dystem Dynamic Performance*; IEEE Press: Piscataway, NJ, USA, 2003.

31. Zhao, Q.; Jiang, J. Robust SVC controller design for improving power system damping. *IEEE Trans. Power Syst.* **1995**, *10*, 1927–1932. [CrossRef]

32. Noroozian, M.; Ghandhari, M.; Andersson, G.; Gronquist, J.; Hiskens, I. A robust control strategy for shunt and series reactive compensators to damp electromechanical oscillations. *IEEE Trans. Power Deliv.* **2001**, *16*, 812–817. [CrossRef]

33. Liu, Q.; Vittal, V.; Elia, N. LPV supplementary damping controller design for a thyristor controlled series capacitor (TCSC) device. *IEEE Trans. Power Syst.* **2006**, *21*, 1242–1249. [CrossRef]

34. Zhang, S.; Vittal, V. Design of Wide-Area Power System Damping Controllers Resilient to Communication Failures. *IEEE Trans. Power Syst.* **2013**, *28*, 4292–4300. [CrossRef]

35. Moeini, A.; Kamwa, I. Analytical Concepts for Reactive Power Based Primary Frequency Control in Power Systems. *IEEE Trans. Power Syst.* **2016**, *31*, 4217–4230. [CrossRef]

36. Li, Y.; Li, Y.W. Power Management of Inverter Interfaced Autonomous Microgrid Based on Virtual Frequency-Voltage Frame. *IEEE Trans. Smart Grid* **2011**, *2*, 30–40. [CrossRef]

37. Sun, M.; Jia, Q. A Novel Frequency Regulation Strategy for Single-Stage Grid-Connected PV Generation. In Proceedings of the 2018 2nd IEEE Conference on Energy Internet and Energy System Integration (EI2), Bejing, China, 20–22 October 2018; pp. 1–6. [CrossRef]
38. Dspace Manual. Available online: http://www.dspace.com. (accessed on 1 December 2019).
39. Typhoon Hil manual. Available online: https://www.typhoon-hil.com. (accessed on 1 December 2019).
40. Ren, W.; Steurer, M.; Baldwin, T.L. Improve the Stability and the Accuracy of Power Hardware-in-the-Loop Simulation by Selecting Appropriate Interface Algorithms. *IEEE Trans. Ind. Appl.* **2008**, *44*, 1286–1294. [CrossRef]
41. Wang, J.; Song, Y.; Li, W.; Guo, J.; Monti, A. Development of a Universal Platform for Hardware In-the-Loop Testing of Microgrids. *IEEE Trans. Ind. Inf.* **2014**, *10*, 2154–2165. [CrossRef]
42. Fitzgerald, A.E. *Electric Machinery*, 6th ed.; McGraw-Hill: New York, NY, USA, 2002.

Article

Alienation Coefficient and Wigner Distribution Function Based Protection Scheme for Hybrid Power System Network with Renewable Energy Penetration

Sheesh Ram Ola [1], Amit Saraswat [1,*], Sunil Kumar Goyal [1], Virendra Sharma [2], Baseem Khan [3], Om Prakash Mahela [4], Hassan Haes Alhelou [5,*] and Pierluigi Siano [6,*]

[1]   Department of Electrical Engineering, Manipal University Jaipur, Rajasthan 303007, India; sheeshola@gmail.com (S.R.O.); sunilkumar.goyal@jaipur.manipal.edu (S.K.G.)
[2]   Department of Electrical Engineering, Arya College of Engineering and Information Technology, Jaipur 302028, India; vsharmakiran@gmail.com
[3]   Department of Electrical Engineering, Hawassa University, Awasa P.O. Box 05, Ethiopia; baseem.khan04@gmail.com
[4]   Power System Planning Division, Rajasthan Rajya Vidhyut Prasaran Nigam Ltd., Jaipur 302005, India; opmahela@gmail.com
[5]   Department of Electrical Power Engineering, Faculty of Mechanical and Electrical Engineering, Tishreen University, 2230 Lattakia, Syria
[6]   Department of Management & Innovation Systems, University of Salerno, 84084 Fisciano (SA), Italy
*   Correspondence: amit.saraswat@jaipur.manipal.edu (A.S.); alhelou@tishreen.edu.sy (H.H.A.); psiano@unisa.it (P.S.)

Received: 4 February 2020; Accepted: 25 February 2020; Published: 2 March 2020

Abstract: The rapid growth of grid integrated renewable energy (RE) sources resulted in development of the hybrid grids. Variable nature of RE generation resulted in problems related to the power quality (PQ), power system reliability, and adversely affects the protection relay operation. High penetration of RE to the utility grid is achieved using multi-tapped lines for integrating the wind and solar energy and also to supply loads. This created considerable challenges for power system protection. To overcome these challenges, an algorithm is introduced in this paper for providing protection to the hybrid grid with high RE penetration level. All types of fault were identified using a fault index (FI), which is based on both the voltage and current features. This FI is computed using element to element multiplication of current-based Wigner distribution index (WD-index) and voltage-based alienation index (ALN-index). Application of the algorithm is generalized by testing the algorithm for the recognition of faults during different scenarios such as fault at different locations on hybrid grid, different fault incident angles, fault impedances, sampling frequency, hybrid line consisting of overhead (OH) line and underground (UG) cable sections, and presence of noise. The algorithm is successfully tested for discriminating the switching events from the faulty events. Faults were classified using the number of faulty phases recognized using FI. A ground fault index (GFI) computed using the zero sequence current-based WD-index is also introduced for differentiating double phase and double phase to ground faults. The algorithm is validated using IEEE-13 nodes test network modelled as hybrid grid by integrating wind and solar energy plants. Performance of algorithm is effectively established by comparing with the discrete wavelet transform (DWT) and Stockwell transform based protection schemes.

Keywords: alienation coefficient; hybrid power system network; protection; power system fault; solar energy; wind energy; Wigner distribution function

---

1. Introduction

Adverse environmental impacts of fossil fuel based power plants have forced the utilities to integrate clean energy with the grid in order to meet future energy demands. Fast development of renewable technologies and government incentives to reduce carbon footprints have motivated the utilities to switch from the conventional power plants to the renewable energy (RE) generation sources [1]. This is achieved by forming the hybrid grid with multi-tapped transmission and sub-transmission lines to supply the loads and integrate RE sources such as wind and solar power plants. Formation of hybrid grids using multi-tapped lines provide economic solutions for RE integration to the grid; however it creates protection challenges due to variable nature of RE generation and bidirectional flow of power in the lines. This resulted in the requirement of new protection schemes, which can be deployed in the recent structure of hybrid grids for effective protection. These techniques must be independent of direction of power flow, unbalanced nature of loads, fault current and variable generation [2]. This can be effectively achieved by the use of machine learning and signal processing techniques. Fang et al. [3], proposed an improved distance relay scheme using time delay and zero-sequence impedance for the grid to which RE sources are integrated. This scheme has high reliability compared to the conventional relays and reduced risk of malfunction. A detailed study of challenges associated with the protection of grid integrated distributed generation (DG) and adaptive protection schemes for these systems are presented in [4]. A detailed study related to the application of signal processing techniques and intelligent methods such as artificial neural network (ANN), fuzzy set theory (FST), and expert system (ES) in the field of protection of DG sources integrated power system is presented in [5]. A sensor based fault detection isolation scheme for the grid, interfaced with RE sources and electric vehicles (EV) is introduced by authors in [6]. This can effectively be deployed in eleven order multi-area smart dynamic power system interfacing RE and EV. In [7], authors introduced an algorithm based on the fast recursive discrete Fourier transform (FRDFT) for the protection of distribution system, integrated with DG. This is a novel, fast and adaptive relay technique for relay systems, which is effective for obtaining the optimal protection settings when system conditions are continuously changing. Application of the syntactic methods for identification of power system signals by measuring the parameters is reported in [8,9]. This method has the capability to provide syntactic and semantic information simultaneously. Nieto et al. [10], introduced a detailed study for improvement of quality of power in grid with integration of energy storage systems. This study helps to differentiate the power quality disturbances from the disturbances associated with the faulty events. In [11], a current-based transmission line protection scheme using Wigner distribution function and alienation coefficient is introduced. However, this scheme fails to provided protection to hybrid grids in the presence of both wind and solar energy penetration. Discrete wavelet transform was implemented for the identification of wind and solar energy penetration into the utility grid [12,13]. This method has the disadvantage of generating false tripping signals in the presence of high noise level. Stockwell transform was implemented for the identification of wind and solar energy penetration into the utility grid [14,15]. This method overcomes the demerits of discrete wavelet transform (DWT) scheme; however, it has the disadvantage of slow speed of protection scheme due to large data involved in the extraction of features from voltage and current signals. Therefore, new protection scheme is designed to provide effective protection against various faults to the hybrid grid with high RE penetration level. This protection scheme has merits, such as fast response (recognition of fault in a time duration which is less than quarter cycle) and performance is not affected by presence of high level of noise. Following are main contributions of the paper:

- An algorithm supported by the features of both voltage and current is introduced in this paper, which is effective to provide protection for the hybrid grid with RE penetration.
- A W-index is computed by processing the current signals using Wigner distribution function (WDF). Also an ALN-index is computed using sample-based alienation coefficients of voltage signals. Furthermore, fault index (FI) is computed from these W-index and ALN-index, which is effective for recognition of all the fault types incident on hybrid grid.

- The performance of the algorithm is not affected by the presence of high level of noise.
- The proposed algorithm is effective to provide high speed protection of hybrid grid. Faults were detected within a time less than $(1/10)$th cycle.
- The algorithm is effective in discriminating switching transients from the faulty transients.
- The fault type was recognized using the number of faulty phases and identified using FI. A GFI, based on the processing of zero sequence currents using WDF is effective in differentiating the double phase (2P) and double phase to ground (2PG) faults.
- This algorithm works effectively and efficiently for the identification of faults on the hybrid grid during various operating scenarios such as fault location, various sampling frequencies, variations in the fault impedance, fault incidence angle and hybrid combination of the OH line and UG cable.

Nine sections are used to arrange the contents in the article. A brief introduction to carry out research on the selected topic and main contributions is included in Section 1. Section 2 describes the hybrid grid test network incorporated with RE generators. The proposed algorithm implemented for the protection of hybrid grid is described in Section 3. Section 4 demonstrates the simulation results related to fault identification in the hybrid grid. This section also includes the simulation results to highlight the requirements of protection scheme for the hybrid grid. Fault classification results are included in Section 5. Section 6 discusses the results of various case studies. Results for the discrimination of switching events from the faulty events are discussed in Section 7. A brief comment on the results and performance comparison of the proposed protection scheme with other existed schemes in the presence of RE is detailed in Section 8, followed by the conclusions (Section 9).

2. The Proposed Test System

IEEE-13 node test network is modelled as hybrid grid by incorporating the wind and solar photovoltaic (PV) generators. As specified by IEEE, the capacity of this network is equal to $5MVA$ and operated at frequency of 60 Hz and two voltage levels of 0.48 kV and 4.16 kV [16,17]. Doubly fed induction generator-based wind plant (WG) of capacity 1.5 MW is integrated at node 680 of test system using transformer TRFW. Also, PV plant of capacity 1 MW is integrated to test network at node 680 using a transformer TRFS and an overhead line (5 km length) as described in Figure 1. Positive and zero sequence resistances of this line are 0.1153 and 0.413 Ω/km respectively. Positive and zero sequence inductances of the same line are $1.05e^{-3}$ and $3.32e^{-3}$ H/km respectively. Positive and zero sequence capacitances of the same line are $11.33e^{-9}$ and $5.01e^{-9}$ F/km respectively. Since, wind and solar PV plants are integrated at node 680, therefore this node is considered to be a point of common coupling (PCC).

The configurations of overhead (OH) lines and underground (UG) cables are considered same as that provided in the original data as illustrated in Table 1. Impedance matrix (Z_{601}) of the OH lines (configuration 601) is provided by the equation (1) where all the impedances are expressed in Ω/km. Impedance matrix (Z_{606}) of UG cables (configuration 606) is provided by equation (2) where all impedances are expressed in Ω/km. Positive and zero sequence capacitance magnitudes of 1.57199 nF/km and $1.3398nF/km$, respectively are used for OH lines. Similarly, the parameters for UG cables (configuration 606) are taken as 15.96979 µF/km. All loads are assumed as 3ϕ balanced as detailed in Table 2 [18,19].

$$Z_{601} = \begin{bmatrix} 0.2153 + j0.6325 & 0.0969 + j0.3117 & 0.0982 + j0.2632 \\ 0.0969 + j0.3117 & 0.2097 + j0.6511 & 0.0954 + j0.2392 \\ 0.0982 + j0.2632 & 0.0954 + j0.2392 & 0.2121 + j0.6430 \end{bmatrix} \tag{1}$$

$$Z_{606} = \begin{bmatrix} 0.4960 + j0.2773 & 0.1983 + j0.0204 & 0.1770 + j0.0089 \\ 0.1983 + j0.0204 & 0.4903 + j0.2511 & 0.1983 + j0.0204 \\ 0.1770 + j0.0089 & 0.1983 + j0.0204 & 0.4960 + j0.2773 \end{bmatrix} \tag{2}$$

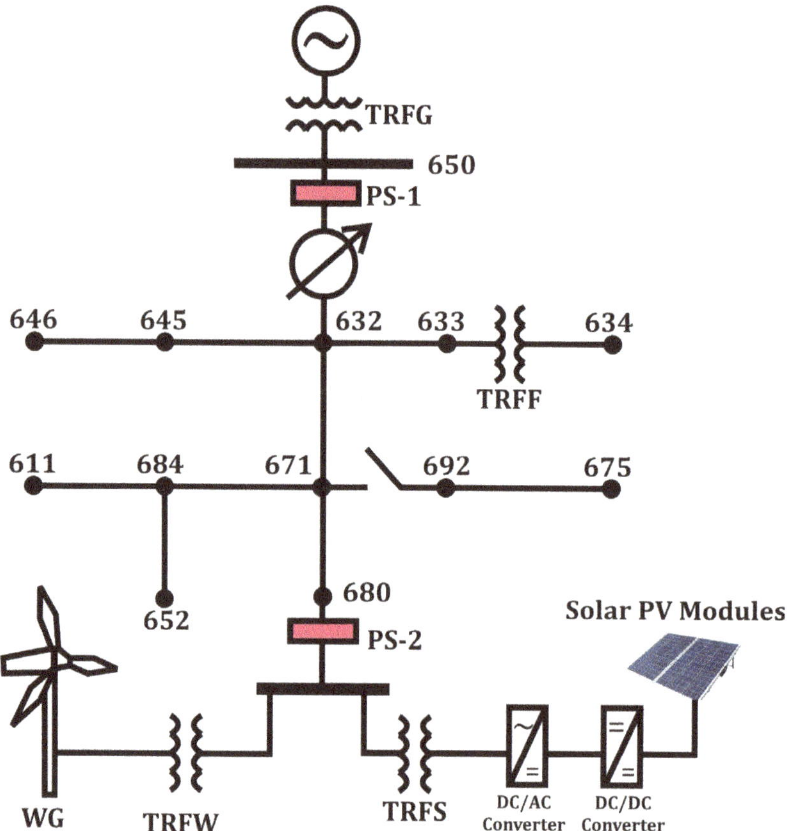

Figure 1. Hybrid power system network incorporating wind and solar energy plants.

Table 1. Feeder Data of Hybrid Grid.

Node A	Node B	Length of Feeder (m)	Configuration of Feeder
632	645	152.40	601
632	633	152.40	601
633	634	0	XFM
645	646	91.440	601
650	632	609.60	601
684	652	243.84	606
632	671	609.60	601
671	684	680.00	601
671	692	0	Switch
684	611	91.44	601
692	675	152.40	606

Table 2. Loading Data of Hybrid Grid.

Nodes	Load Model	Load		Capacitor Banks
		kW	kVAr	(kVAr)
634	Y-PQ	400	290	
645	Y-PQ	170	125	
646	Y-PQ	230	132	
652	Y-PQ	128	86	
671	Y-PQ	1155	660	
675	Y-PQ	843	462	600
692	Y-PQ	170	151	
611	Y-PQ	170	80	100
632–671	Y-PQ	200	116	

Test network of the hybrid grid with RE penetration is integrated to the network of utility grid using a transformer designated as TRFG. A transformer designated as TRFF is used as interconnecting transformer (ICT) between the nodes 633 and 634 of hybrid grid. Parameters of transformers used in the hybrid grid are provided in Table 3.

Table 3. Transformer Parameters of Hybrid Grid with RE Penetration.

Transformer	MVA	kV	kV	HV Winding		LV Winding	
		High	Low	$R(\Omega)$	$X(\Omega)$	$R(\Omega)$	$X(\Omega)$
TRFG	10	115	4.16	29.095	211.60	0.1142	0.8306
TRFF	5	4.16	0.48	0.3807	2.7688	0.0510	0.0042
TRFS	1	4.16	0.260	0.1730	195.70	0.0007	0.7645
TRFW	5	4.16	0.48	0.3807	2.7688	0.0510	0.0042

Protection schemes PS-1 and PS-2 are installed at nodes 650 and 680 (PCC), respectively of the hybrid grid. These schemes will be used to disconnect the test hybrid grid and RE generators from the network of utility grid during the faulty events. Voltage and currents are continuously processed using the proposed algorithm and tripping signals are given to the respective circuit breaker (CB) if fault is detected. Protection schemes may be installed at any node of the hybrid grid by looking towards the actual requirement of protection. This algorithm will work well for all the locations of hybrid grid using a suitable weight factor for the proposed fault index. Results are discussed in detail for the protection scheme PS-1.

2.1. Wind Energy Conversion System

The wind energy conversion system (WECS) is comprised of a wind turbine, a doubly fed induction generator (DFIG) and a converter system. Kinetic energy of the wind is converted into mechanical energy of wind turbine shaft, which is converted into the electrical energy using the DFIG, coupled mechanically to the shaft of wind turbine. WECS consists of DFIG, with the capacity of 1.5 MW and operating at 60 Hz frequency with output voltage of $575V$ [20]. Rated wind speed equal to 11 m/s is considered and generators have the following data: H (inertia constant)= 0.685 s, $Rs = 0.023$ pu, $Ls = 0.18$ pu, $Rr = 0.016$ pu, $Lr = 0.016$ pu, $Lm = 2.9$ pu. Wind turbine, DFIG and their control systems were modelled using the parameters reported in [21,22].

2.2. Solar Photovoltaic System

The solar photovoltaic (PV) plant consists of components such as solar PV plates, boost converter, maximum power point tracking (MPPT) system, inverter, transformer, grid coupling inductor and capacitor. The solar panel where actual power is generated (solar to electrical energy transformation) comprises of series and parallel combinations of solar cells. Solar cell is fabricated in a thin layer of semiconductor in the form of p-n junction (p-n diode), which has the same operational characteristics as the p-n junction diode. Characteristics are dependent on the quantity of solar radiations and temperature of PV plates. A single diode equivalent model of solar PV cell with parallel and series resistances is used in this study and illustrated in Figure 2. This is simple in nature and has sufficient accuracy [23]. This solar PV system is interfaced to the test system to design a hybrid grid.

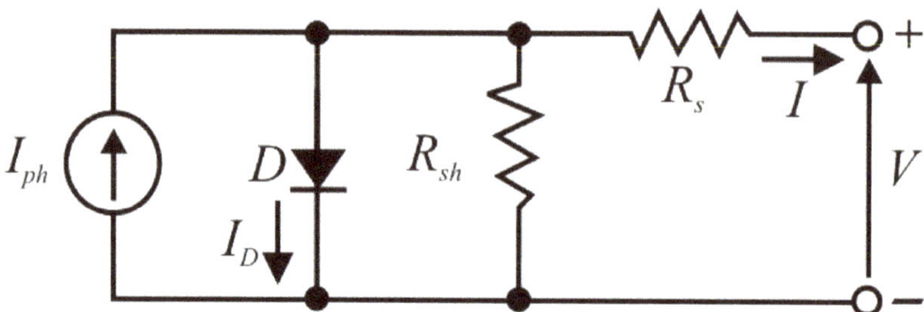

Figure 2. Solar cell modelled using single diode equivalent.

The solar PV plant of capacity 1 MW is formed using the ten units (each with capacity 100 kW) and integrated in parallel. The output voltage (V) of a solar cell (single-diode equivalent circuit) is related to the output current (I) as per following relation [24].

$$I = I_{ph} - I_0 \left\{ exp \left[\frac{q(V + IR_s)}{AkT} \right] - 1 \right\} - \left(\frac{V + IR_s}{R_{sh}} \right) \tag{3}$$

where I_{ph}: photo-current of solar PV cell, I_0: saturation current of PV cell, A: curve fitting factor of PV cell, R_{sh}: shunt resistance of PV cell, R_s: series resistance of PV cell, q: electronic charge, and k: Boltzmann constant.

Magnitude of R_{sh} is infinite at short circuit conditions. At this condition slope of I-V characteristics tends to zero [24]. Hence, I_{ph} is equal to short circuit current (I_{sc}) [23]. For a PV array organized in N_p parallel and N_s series connected solar cells, the current is expressed as below.

$$I = N_p I_{sc} - N_p I_0 \left\{ exp \left[\frac{q(V + I(N_s/N_p)R_s)}{N_s AkT} \right] - 1 \right\} \tag{4}$$

Parameters simulated in this study for each module (at standard test condition) are as $V_{oc} = 64.2V$, $I_{sc} = 5.96A$, $V_{mp} = 54.7V$, $I_{mp} = 5.58A$, $R_s = 0.037998\Omega$, $R_{sh} = 993.51\Omega$, $I_0 = 1.1753e^{-8}A$, diode quality factor $Q_d = 1.3$, and $I_{ph} = 5.9602A$ [25]. The V_{mp} and I_{mp} are respectively the voltage and current at point of maximum power tracking.

3. The Proposed Algorithm for the Protection Scheme

The algorithm for the recognition of faults in the hybrid power system, incorporated with RE sources, is based on features that are extracted from both the voltage and current signals. The proposed algorithm is implemented in two steps, where in first step faults are detected and in second step, faults are classified. Furthermore, detection of the faults is also performed in two steps, where current-based Wigner distribution index (WD-index) and voltage-based alienation index (ALN-index) are evaluated. The WD-index and ALN-index are multiplied to obtain the proposed fault index (FI), which is found to be effective for the discrimination of faulty events from healthy condition as well as faulty phase from healthy phase. The FI corresponding to the faulty phase will have a higher magnitude compared to the set threshold magnitude (TM) and it has a lower value than TM corresponding to the healthy phase during all conditions. The algorithm was tested on 30 sets of data for each fault type to set a threshold value of 5000. The data set is obtained by changing the parameters including fault impedance, incidence angle of fault, location of faults at various nodes of hybrid grid, presence of noise on both voltage and current signals etc. Classification of the faults is achieved by estimating the number of faulty phases. However, a ground fault index (GFI) using Wigner distribution function based decomposition of negative sequence current signals is introduced for discriminating double phase fault, with and without the involvement of ground. Deviations in the patterns of current and voltage waveforms during faulty events were used for the identification of faults using proposed algorithm; therefore there is no requirement for parameter normalization. The performance of the algorithm will be affected by the power network configuration and penetration level of the RE. However, the algorithm can be used in the different network configurations and RE penetration levels by changing the threshold magnitude. All steps of algorithm are described with the help of flow chart as illustrated in Figure 3. FI, GFI, current-based WD-index and voltage-based ALN-index are described below. The study is performed in MATLAB/Simulink 2017a software environment on a laptop computer which has a 64 bit operating system, RAM of 4 GB, and Intel (I) Core(TM) i5-3230M CPU@2.60 GHz processor.

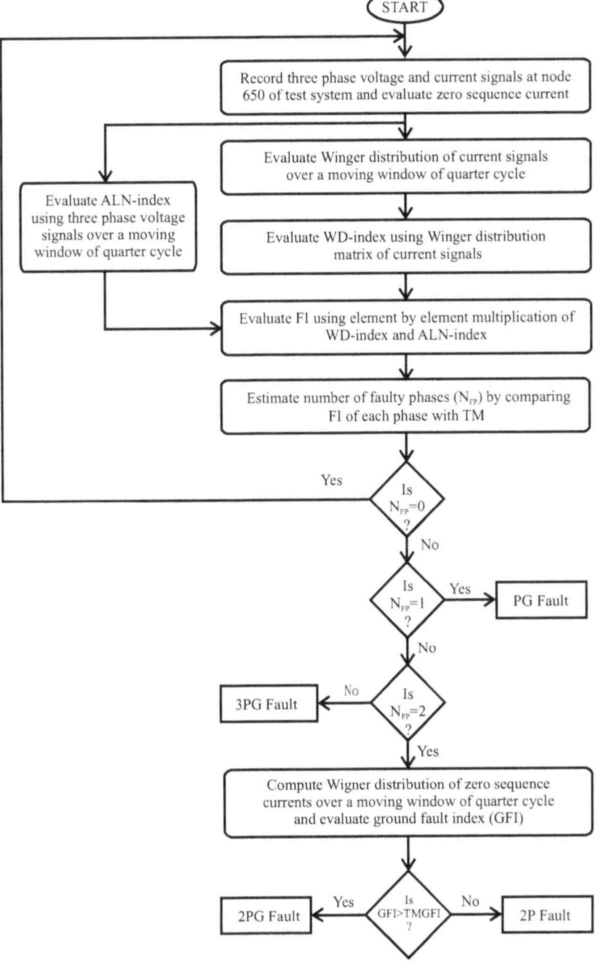

Figure 3. The algorithm for the recognition of faults on the hybrid power system incorporating renewable energy.

3.1. Current Based Wigner Distribution Index

Wigner distribution function (WDF) is a phase space distribution function. This is effective in description of signal in a space and frequency at the same time. This can be considered to be a local frequency spectrum of the signal. This can be used for processing of both deterministic and stochastic signals. The WDF is effective in providing local frequency spectrum of the signal. Hence, it is most suitable for analysis of the faulty transients to identify the faulty events. The current signals captured at node 650 are processed using Wigner distribution function over a quarter cycle with a sampling frequency (SF) of 3.84 kHz. Window moving by one sample step is used for the continuous computation of Wigner distribution. Absolute magnitude of the output is evaluated and designated as WD-index. Energy density of current signal is used by the WD-index for estimation of the faults. This is achieved by Bilinear analysis of current signal $I(t)$ (in time domain) twice. The WD-index has the advantage of high concentration of energy as well as high resolution of time-frequency [26]. Following relation effectively illustrates the evaluation of WD-index by processing the current signal ($I(t)$) [27].

$$WDindex = \int_{-\infty}^{\infty} I(t + \frac{\tau}{2})I^*(t + \frac{\tau}{2})e^{-j\omega\tau}d\tau \tag{5}$$

here representations of symbols are as follows t: time (sliding variable); ω: signal angular frequency; τ: time domain based signal function.

3.2. Voltage Based Alienation Index

Alienation index is computed using sample-based alienation coefficients of voltage signals at a SF of 3.84 kHz and designated as ALN-index. This index is evaluated using the correlation coefficient (r) between voltage magnitudes at two different time instants as detailed below.

$$ALNindex = 1 - r^2 \tag{6}$$

Here, r is the correlation coefficient between voltage variables x and y and can be expressed as detailed below.

$$r = \frac{N_s \sum xy - (\sum x)(\sum y)}{\sqrt{[N_s \sum x^2 - (\sum x)^2][N_s \sum y^2 - (\sum y)^2]}} \tag{7}$$

here N_s is the numbers considered in a cycle (in this study N_s=64 is considered), x is the voltage samples measured at time t_0, y is the voltage samples measured at $-T + t_0$ time where T is time period of voltage signal [28,29]. The ALN-index is evaluated with the help of moving window technique for the samples of quarter cycle. Implementation of this index is based on the comparison of data of quarter cycle considered with the data of previous quarter cycle using a moving window (one sample step). This index is effective in reducing the fault detection time because it has the merit of sharp change at the time of fault incidence. This makes the protection scheme fast.

3.3. Fault Index Based on Voltage and Current Features

A fault index (FI), based on the features of both voltage and current is introduced for recognition of faults in the hybrid power system with RE penetration. This index is computed using element to element multiplication of ALN-index and WD-index as detailed in the following relation.

$$FI = (ALNindex) \times (WDindex) \tag{8}$$

The proposed FI effectively identifies the different nature of faults in the hybrid power network with RE penetration by good accuracy and minimum time. This is achieved by comparing the magnitude of FI with pre-set threshold magnitude (TM). Individual application of ALN-index will not detect any type of fault due to its same value for all phases for all the events. However, application of

WD-index individually will recognize the faults but time of fault identification will be low, which will reduce the protection speed. Hence, FI combines the merits of both ALN-index and WD-index.

3.4. Ground Fault Index

Faults 2P and 2PG cannot be differentiated from each other based on the number of faulty phases. This is achieved by introducing the ground fault index (GFI), which is obtained by processing the zero sequence component of the current signal sampled at the frequency of 3.84 kHz using Wigner distribution function as illustrated in Equation (5). A threshold magnitude for ground fault index (TMGFI) is set equal to 2000 for differentiating the 2P and 2PG faults. Magnitude of GFI greater than TMGFI will indicate the involvement of ground in the faulty event. Hence, values of GFI greater than TMGFI detects the 2PG fault. However, values low compared to the TMGFI detects the 2P fault.

4. Faulty Event Recognition: Simulation Results

This section firstly elaborates the impacts of faulty events on performance of the hybrid grid and requirement of protection schemes. Results of recognition of the faults with a different nature and incidents in the hybrid power network with RE penetration are described and discussed in this section Faults of different types are simulated at node 646 of test system. The current and voltage are recorded at the location of protection scheme (PS-1) at node 650.

4.1. Impacts of Faulty Events on the Performance of Hybrid Grid and Requirement of Protection Schemes

To investigate the impacts of faulty events on the performance of hybrid grid, root mean square (RMS) voltage at the node 650, frequency, power injected by solar PV system and wind power plant into the hybrid grid during the event of PG (phase A) fault at node 646 at 6th cycle are illustrated in Figure 4a–d, respectively. It is observed from Figure 4a that RMS voltage decreases during faulty event whereas small deviations are observed in the frequency for short duration at the time of fault incidence as depicted in Figure 4b. Power supplied by the solar PV and wind generators reduced as depicted in Figure 4c,d, respectively. Hence, performance of the hybrid grid will be affected adversely, if faulty events persist. Therefore, suitable protection scheme needs to be investigated and designed to isolate the faulty section of the hybrid grid.

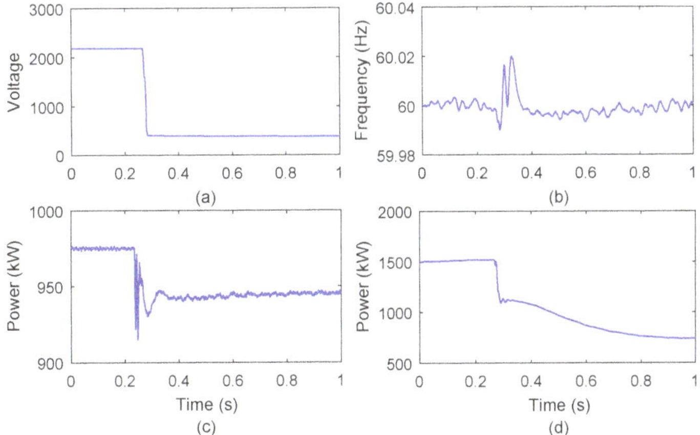

Figure 4. Phase to ground fault at the node 646 of hybrid power system network (**a**) root mean square voltage (**b**) frequency (**c**) power injected by solar PV system into the hybrid grid (**d**) power injected by wind power plant into the hybrid grid.

4.2. Phase to Ground Fault

The phase to ground (PG) fault is simulated on phase A at 0.1 s at node 646. Current and voltage signals captured at node 650 for the period of 0.2 s (12 cycles) are detailed in Figure 5a,b, respectively. Current signals are processed using Wigner distribution function (WDF) and WD-index is computed, which is described in Figure 5c. It is observed that WD-index corresponding to phase A has a high magnitude after incidence of PG fault. However, this index, corresponding to phases B and C, has values comparable to the pre-fault values. The ALN-index is computed from the voltage signals and described in Figure 5d. It is concluded that the ALN-index corresponding to all phases sharply increases just after the incidence of PG fault.

Figure 5e describes the FI corresponding to all the phases during the event of PG fault. It can be inferred that FI corresponding to the faulty phase (phase A) has a higher magnitude compared to TM after the incidence of PG fault. However, this FI corresponding to healthy phases (phases B and C) has a lower magnitude compared to TM. Hence, the algorithm is found to be effective for the identification of PG fault, discriminating the healthy and faulty phases. High resolution of FI is illustrated in Figure 5f. It is observed that FI corresponding to phase A rises and crosses the TM after 6×10^{-4} s, whereas the FI corresponding to healthy phases B and C remains below the threshold. Hence, the PG fault was detected effectively in time duration, equal to 3.6% of the total time of the cycle.

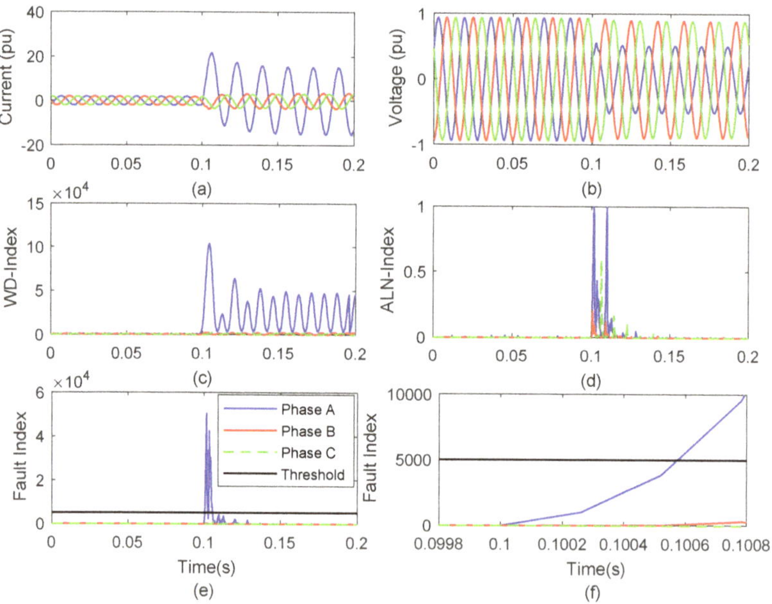

Figure 5. Recognition of PG fault incident at the node 646 of hybrid test system (**a**) current waveform (**b**) voltage waveform (**c**) WD-index (**d**) ALN-index (**e**) FI (**f**) plot to compute fault recognition time.

4.3. Double Phase Fault

The double phase (2P) fault is simulated on phases A and B at 0.1 s at node 646. Current and voltage signals recorded at node 650 for the period of 0.2 s (12 cycles) are illustrated in Figure 6a,b respectively. Current signals are processed using WDF and WD-index is computed which is detailed in Figure 6c. It is observed that WD-index corresponding to phases A and B has a high magnitude after incidence of 2P fault. However, this index corresponding to the phase C has values comparable to the pre-fault values. The ALN-index is computed from the voltage signals and detailed in Figure 6d.

It is inferred that ALN-index corresponding to all the phases sharply increases just after the incidence of 2P fault.

Figure 6e details the FI corresponding to all phases during the event of 2P fault. It is seen that FI corresponding to faulty phases (phases A and B) has a higher magnitude compared to TM, after the incidence of 2P fault. However, FI corresponding to healthy phase (phase C) has a lower magnitude as compared to TM. Hence, the algorithm is found to be effective for the identification of 2P fault and the discrimination of healthy and faulty phases. High resolution of FI is illustrated in Figure 6f. It is observed that FI corresponding to phases A and B rises and cross the TM after 7×10^{-5} s and 4×10^{-5} s, respectively, which are equal to 0.42% and 0.24% of the total time of a cycle in the same order. FI corresponding to healthy phase C remains below the threshold. Hence, the 2P fault was detected effectively in time duration equal to 0.42% of the total time of a cycle.

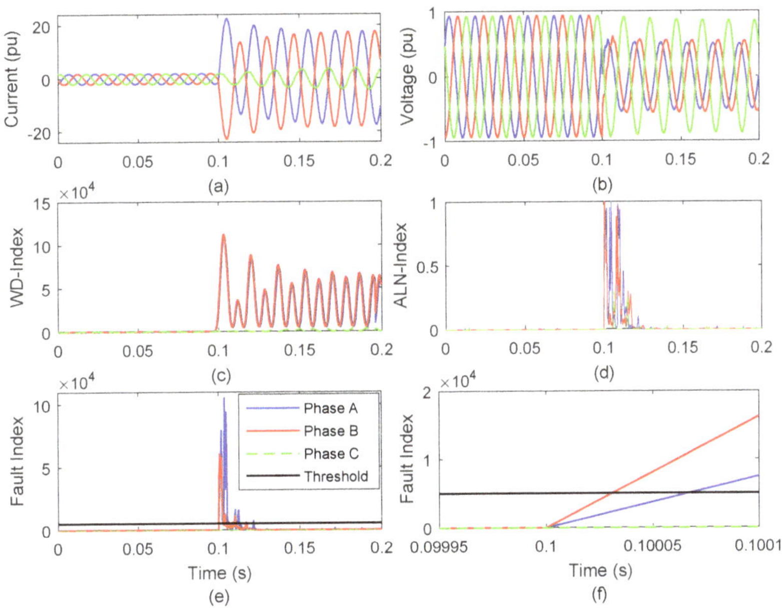

Figure 6. Recognition of 2P fault incident at node 646 of hybrid test system (**a**) current waveform (**b**) voltage waveform (**c**) WD-index (**d**) ALN-index (**e**) FI (**f**) plot to compute fault recognition time.

4.4. Double Phase to Ground Fault

The double phase to ground (2PG) fault is simulated on phases A and B at 0.1 s at node 646. Current and voltage signals recorded at node 650 for the period of 0.2 s (12 cycles) are described in Figure 7a,b, respectively. Current signals are processed using WDF and WD-index is computed, which is detailed in Figure 7c. This is observed that WD-index corresponding to phases A and B has a high magnitude after incidence of 2PG fault. However, this index corresponding to the phase C has values comparable to the pre-fault values. The ALN-index is computed from the voltage signals and detailed in Figure 7d. This is inferred that ALN-index corresponding to all the phases, sharply increases just after incidence of 2PG fault.

Figure 7e details the FI corresponding to all the phases, during the event of 2PG fault. It is seen that FI corresponding to faulty phases (phases A and B) has a higher magnitude compared to TM after the incidence of 2PG fault. However, FI corresponding to the healthy phase (phase C) has a lower magnitude as compared to TM. Hence, the algorithm is found to be effective for the identification of 2PG fault and for the discrimination of the healthy and faulty phases. High resolution of FI is illustrated in Figure 7f. It is observed that FI corresponding to phases A and B rises and cross the TM after 8×10^{-5} s and 3×10^{-5} s, respectively, which are equal to 0.48% and 0.18% of total time of the cycle in same order. FI corresponding to the healthy phase C remains below the threshold. Hence, the 2PG fault was detected effectively in time equal to 0.48% of the total time duration of the cycle.

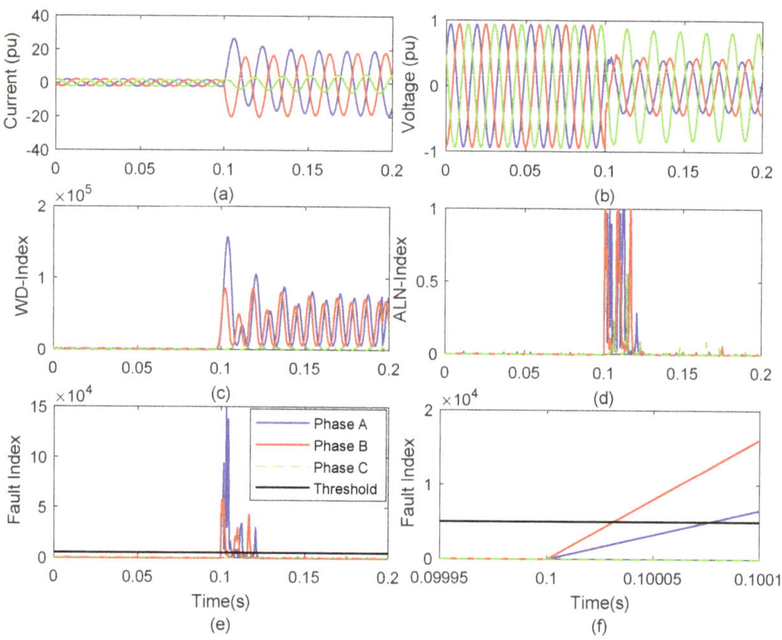

Figure 7. Recognition of 2PG fault incident at node 646 of hybrid test system (**a**) current waveform (**b**) voltage waveform (**c**) WD-index (**d**) ALN-index (**e**) FI (**f**) plot to compute fault recognition time.

4.5. Three Phase to Ground Fault

The three phase to ground (3PG) fault is simulated at 0.1 s at node 646. Current and voltage signals recorded at node 650 for the period of 0.2 s (12 cycles) are described in Figure 8a,b, respectively. Current signals are processed using Wigner distribution function (WDF) and WD-index is computed which is detailed in Figure 8c. It is observed that WD-index corresponding to all phases has a high magnitude after the incidence of 3PG fault. The ALN-index is computed from the voltage signals and detailed in Figure 8d. It is concluded that ALN-index corresponding to all phases sharply increases just after incidence of 3PG fault.

Figure 8e details the FI corresponding to all the phases during the event of 3PG fault. It is seen that FI corresponding to all the phases has a higher magnitude compared to TM after the incidence of 3PG fault, which indicates that all the phases are faulty in nature. Hence, the algorithm is found to be effective for the identification of 3PG fault. High resolution of FI is illustrated in Figure 8f. This is observed that FI corresponding to phases A, B, and C rises and cross the TM after 1.7×10^{-4} s, 3×10^{-5} s and 7×10^{5} s, respectively, which are equal to 1.02%, 0.18% and 0.42% of the total time of a cycle in same order. Hence, the 3PG fault was detected effectively in a time duration equal to 1.02% of the total time of a cycle.

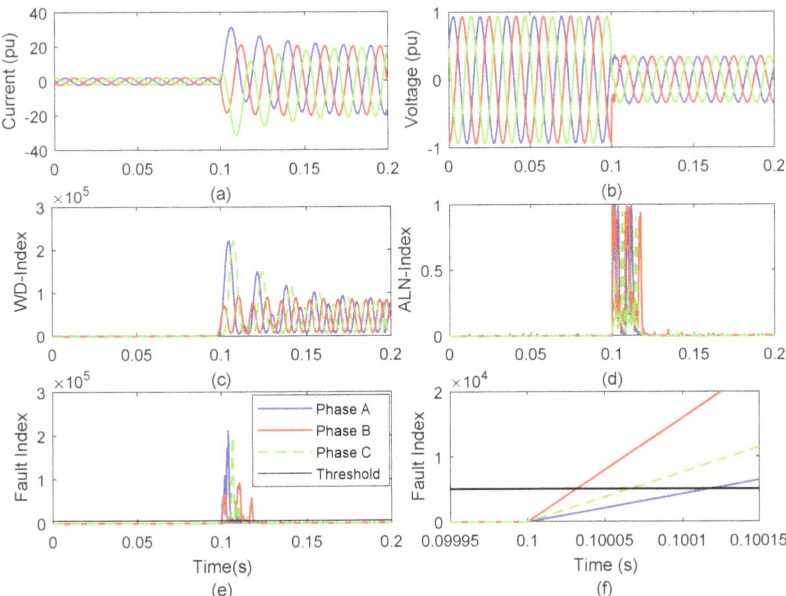

Figure 8. Recognition of 3PG fault incident at node 646 of hybrid test system (**a**) current waveform (**b**) voltage waveform (**c**) WD-index (**d**) ALN-index (**e**) FI (**f**) plot to compute fault recognition time.

Therefore, it is established that all types of faults including PG, 2P, 2PG, and 3PG in the network of the hybrid grid incorporated with RE were identified effectively by using proposed algorithm, within time duration of $(1/10)$th of a cycle.

5. Fault Classification

Type of fault incident on the hybrid power grid in the presence of RE generation were identified using the algorithm illustrated in Figure 3. The PG fault and 3PG fault were identified based on the number of faulty phases, which are 1 and 3, respectively for the PG and 3PG faults. 2P and 2PG faults are included in the same group using the number of faulty phases, identified based on the proposed FI. These faults were classified using the proposed ground fault index as detailed in Figure 9. It is observed that the magnitude of GFI is higher compared to the TMGFI (2000) for the 2PG fault event. However, during the event of 2P fault, the GFI has values lower than the TMGFI. Hence, all the faults were identified effectively using the proposed algorithm

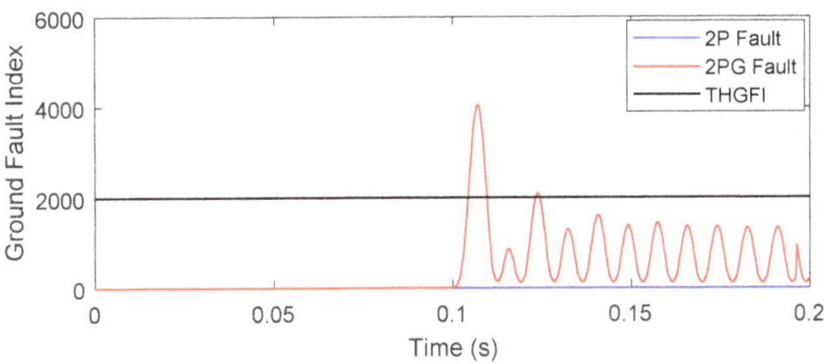

Figure 9. Ground fault index for classification of 2P and 2PG faults.

6. Implementation of the Algorithm to Recognize Faulty Events with Various Scenarios: Case Studies

To generalize the implementation of proposed protection scheme to provide effective protections for hybrid grid with RE penetration, algorithm was tested for recognition of the faults during various scenarios such as fault location at different nodes of the hybrid grid, variations in fault impedance, variations in fault incidence angle, effect of noise, effect of change in sampling frequency, effect of transformers and effect of hybrid line with underground (UG) cable and overhead (OH) line sections. The algorithm was also tested for identification of the faulty events by measuring voltage and current signals at PCC. Detailed results for recognition of the PG fault during above mentioned scenarios are discussed in following sections. The algorithm is equally applicable for all fault types.

6.1. Fault Location

Incidence of fault on a particular node/line of the hybrid grid affects the nature of faulty transients, which ultimately affects the performance of fault recognition algorithm. To investigate this effect, the algorithm was tested for the faults, simulated at various critical locations of the hybrid grid. FI corresponding to all the phases during the PG fault event at various nodes of hybrid grid are tabulated in Table 4.

Table 4. Fault Index at Different Nodes of Test System with PG Fault.

Test Node	Fault Index Magnitude		
	Phase-A	**Phase-B**	**Phase-C**
646	5.055×10^4	332	1370
632	1.138×10^5	645	1855
634	5.001×10^4	348	998
652	4.956×10^4	1170	439
611	6.216×10^4	1491	1785
671	6.234×10^4	2711	2886
675	6.164×10^4	1529	2001

It can be observed from Table 4 that the FI corresponding to the faulty phase A is higher compared to TM at all locations. However, healthy phases B and C have lower FI compared to TM. Hence, PG faulty is effectively recognized at all nodes of the test hybrid grid with RE penetration. Detailed discussion of PG fault at node 652 and 634 is presented in following subsections.

6.1.1. Fault on Node 652

The algorithm was tested for all the faulty events simulated at node 652 and found to be effective for the recognition of these events. Position of this node is critical because there is hybrid combination of line between this node and node 650 (where current and voltage are recorded), which consists of UG cable and OH line sections. At the junction of UG cable and OH line, reflection of the faulty transient, carrying travelling wave may take place, which might affect the performance of algorithm. Results of PG fault simulated on phase A at 0.1 s on node 652 are included in Figure 10. Current and voltage signals captured on node 650 for a period of 0.2 s (12 cycles) are detailed in Figure 10a,b, respectively. Current signals are processed using WDF and WD-index is computed, which is described in Figure 10c. It is observed that WD-index corresponding to phase A has a high magnitude after the incidence of PG fault. However, this index corresponding to the phases B and C has values comparable to the pre-fault values. The ALN-index is computed from the voltage signals and described in Figure 10d. It is concluded that the ALN-index corresponding to all phases sharply increases just after the incidence of PG fault.

Figure 5e describes the FI corresponding to all the phases during the event of PG fault. It can be inferred that FI corresponding to faulty phase (phase A) has a higher magnitude compared to TM after the incidence of PG fault. However, FI corresponding to healthy phases B and C has a lower magnitude compared to TM. Hence, it is established that algorithm is effective for the identification of PG fault and discrimination of the healthy and faulty phases even when there is hybrid combination of UG cable and OH line sections.

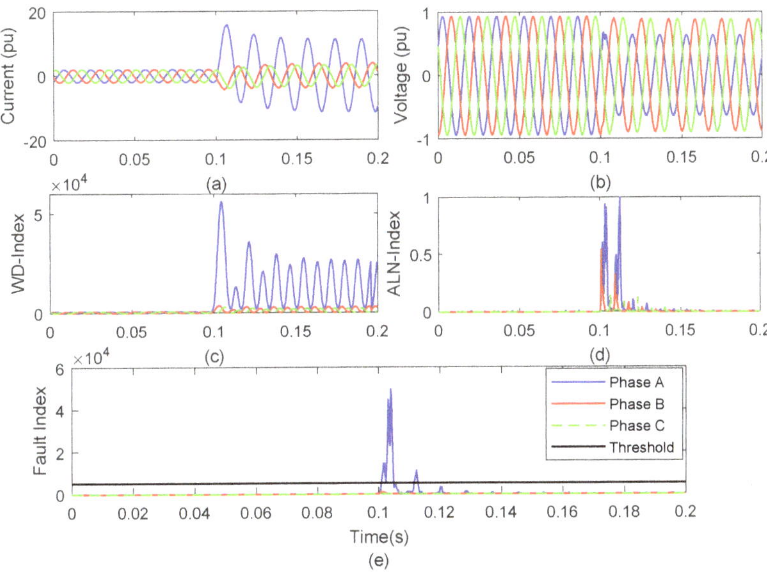

Figure 10. Recognition of PG fault incident on node 652 of hybrid test system (**a**) current waveform (**b**) voltage waveform (**c**) WD-index (**d**) ALN-index (**e**) FI (**f**) plot to compute fault recognition time.

6.1.2. Fault on Node 634

The algorithm was tested for all the faulty events simulated on node 634 and found to be effective for the recognition of these events. This node is considered due to the availability of a transformer (TRFF) between nodes 633 and 634. Presence of transformer in the path of travelling wave carrying faulty transients may affect the performance of algorithm as transformer is a magnetically coupled inductive element. Node 634 is operated at voltage of 0.48 kV and node 633 is operated at the voltage of 4.16 kV. Results for PG fault simulated on phase A at 0.1 s on node 634 are illustrated in Figure 11. Current and voltage signals captured at node 650 for a period of 0.2 s (12 cycles) are detailed in Figure 11a,b, respectively. Current signals are processed using WDF and WD-index is computed, which is described in Figure 11c. It is observed that WD-index corresponding to phase A has a high magnitude after the incidence of PG fault. However, this index corresponding to phases B and C has values comparable to the pre-fault values. The ALN-index is computed from the voltage signals and described in Figure 11d. It is concluded that the ALN-index corresponding to all phases sharply increases just after the incidence of PG fault.

Figure 11e details the FI corresponding to all the phases during the event of PG fault. It can be inferred that FI corresponding to faulty phase (phase A) has a higher magnitude compared to TM after incidence of PG fault. However, FI corresponding to healthy phases B and C has a lower magnitude as compared to TM. Hence, it is established that algorithm is effective for the identification of PG fault and discrimination of the healthy and faulty phases even when there is a transformer between the faulty point and relay location.

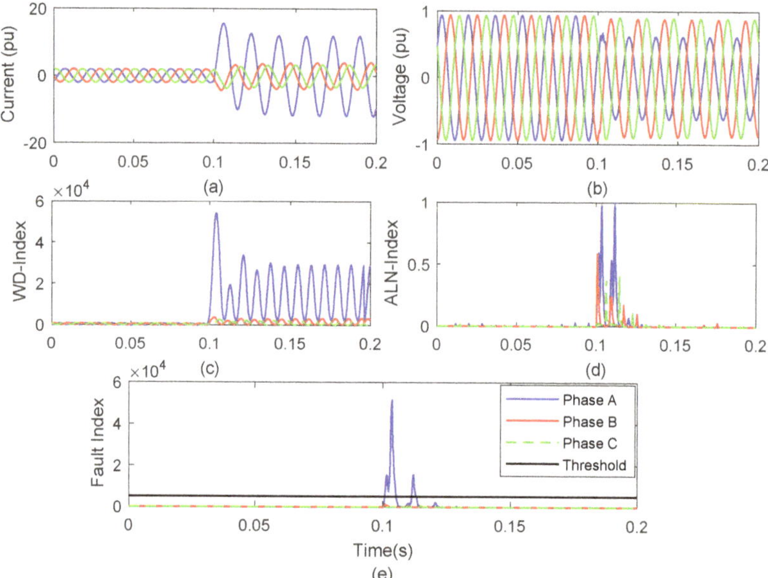

Figure 11. Recognition of PG fault incident on node 634 of hybrid test system (**a**) current waveform (**b**) voltage waveform (**c**) WD-index (**d**) ALN-index (**e**) FI (f) plot to compute fault recognition time.

6.2. Fault Impedance Variations

The nature of fault transients is affected by the fault impedance. Hence, fault impedance might affect the performance of the protection algorithm. In common faulty events, the fault impedance ranges from 0 Ω to 5 Ω. Therefore, the algorithm was tested for the fault impedances of 0 Ω, 5 Ω, 10 Ω, 15 Ω, 20 Ω and 25 Ω for all types of faults. FI magnitude associated with all the phases during the PG fault event incident at node 646 is tabulated in Table 5. It can be observed that FI magnitude is higher compared to TM (5000), corresponds to phase A. Further it is lower as compared to TM, corresponds to phases B and C. It is also observed that FI corresponding to faulty phase decreases with increased fault impedance. Therefore, it is established that PG fault will be effectively detected in hybrid grid with RE penetration and fault impedance up to 25 Ω. Furthermore, the algorithm also works efficiently for all types of faults with fault impedances up to 25 Ω.

Table 5. Fault Index During PG Faulty Event with Different Fault Impedance.

Phase Name	Fault Index Magnitude					
	0Ω	5Ω	10Ω	15Ω	20Ω	25Ω
Phase-A	5.055×10^4	4.601×10^4	3.307×10^4	2.170×10^4	1.003×10^4	8252
Phase-B	332	317	284	253	232	199
Phase-C	1370	1101	923	457	333	316

6.3. Fault Incidence Angle Variations

Fault transients may also be affected by the fault incidence angle (FIA), which may result false tripping indication. Hence, performance of the algorithm was tested by simulating the fault at node 646 with incidence angles of 0°, 30°, 60°, 90°, 120° and 150° for all types of investigated faults. FI magnitude associated with all the phases during the PG fault event incident at node 646 for different angles is tabulated in Table 6. It can be observed that FI magnitude is higher compared to TM (5000), corresponds to phase A and lower compared to TM, corresponds to phases B and C for all types of fault incidence angles. It is also observed that FI corresponding to faulty phase is maximum for FIA of 90°. Therefore, it is established that PG fault will be effectively detected in hybrid grid with RE penetration and all types of fault incidence angle with respect to current and voltage waveforms. Furthermore, the algorithm also works efficiently for all types of faults for different FIA.

Table 6. Fault Index During PG Faulty Event with Different Fault Incidence Angles.

Phase Name	Fault Index Magnitude					
	0°	30°	60°	90°	120°	150°
Phase-A	5.055×10^4	6.848×10^4	4.641×10^4	7.531×10^4	3.875×10^4	2.139×10^4
Phase-B	332	1107	621	1323	1146	704
Phase-C	1370	1244	985	1327	156	75

6.4. Effect of Noise

There is a possibility of interference from external factors on the power system distribution and transmission lines while passing through the terrain, which may superimpose noise on the faulty transients travelling from the faulty point to the location of PS. Noise may affect the performance of the protection schemes resulting in false tripping. To investigate the effect of noise on the performance of algorithm, a noise level of 10 dB signal to noise ratio (SNR) is introduced on both the voltage and current signals. The algorithm was tested for all types of faults whereas the results of PG fault simulated on phase A at 0.1 s on node 646 are illustrated in Figure 12, where superimposed noise can be seen easily. Current and voltage signals captured at node 650 for a period of 0.2 s (12 cycles) are detailed in Figure 12a,b, respectively. Current signals are processed using WDF and WD-index is computed, which is described in Figure 12c. It is observed that the WD-index corresponding to phase A has a higher magnitude after the incidence of PG fault. However, this index corresponding to the phases B and C has values comparable to the pre-fault values. Performance of the WD-index is not affected by noise. The ALN-index is computed from the voltage signals and described in Figure 12d. It is concluded that the ALN-index corresponding to all the phases has a high magnitude throughout time range, indicating that performance of ALN-index will be affected by the noise.

Figure 12 details the FI corresponding to all the phases during the event of PG fault with superimposed noise on both voltage and current signals. It can be inferred that FI corresponding to the faulty phase (phase A) has a higher magnitude compared to TM after the incidence of PG fault. However, FI corresponding to healthy phases B and C has a lower magnitude as compared to TM. Hence, it is established that algorithm is effective for the identification of PG fault and discrimination of the healthy and faulty phases even when there is high level noise (10 dB SNR) superimposed on the current and voltage signals in the hybrid grid in the presence of RE generation. High resolution plot of FI is illustrated in Figure 12f, where it can be seen that FI corresponding to phase A crosses the TM and FI for phases B & C is below the TM.

6.5. Sampling Frequency Variations

To investigate the effect of frequency used for sampling the voltage and current signals on the performance of protection scheme, the algorithm was tested for different sampling frequencies. FI corresponding to the faulty phase A during the event of PG fault is observed and equal to 3.981×10^4, 5.055×10^4, and 5.170×10^4 while using the sampling frequencies of 1.92 kHz, 3.84 kHz and 7.68 kHz, respectively. Therefore, sampling frequency, lower than 3.84 kHz, reduces the peak magnitude of FI, which may also go lower compared to TM. Sampling frequency greater than 3.84 kHz increases the fault detection time due to large size of input data set. Hence, 3.84 kHz is observed to be the optimum SF for the protection scheme.

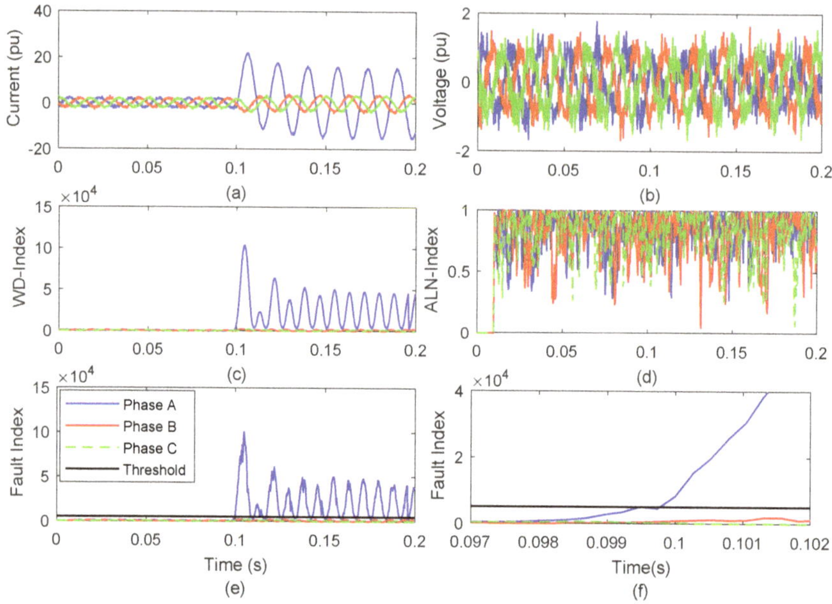

Figure 12. Recognition of PG fault incident on node 652 of hybrid test system in the presence of noise (10dB SNR) (**a**) current waveform (**b**) voltage waveform (**c**) WD-index (**d**) ALN-index (**e**) FI (**f**) plot to compute fault recognition time.

6.6. Recognition of Fault by Recording Voltage and Current Signals at PCC

Performance of the algorithm was tested by recording voltage and current signals at PCC (PS-2 relay location), where RE sources are integrated to test hybrid grid and when PG fault is simulated at node 646 to generalize the applicability of the algorithm. Results of the PG fault, simulated on phase A at 0.1 s at node 646 with measurements at node 680 are illustrated in Figure 13. Current and voltage signals captured at node 680 for a period of 0.2 s (12 cycles) are detailed in Figure 13a,b, respectively. Current signals are processed using WDF and WD-index is computed, which is described in Figure 13c. It is observed that WD-index corresponding to phase A has a higher magnitude after the incidence of PG fault. However, this index corresponding to the phases B and C has values comparable to the pre-fault values. The ALN-index is computed from the voltage signals and described in Figure 13d. It is concluded that the ALN-index corresponding to all phases sharply increases just after the incidence of PG fault.

Figure 13e details the FI corresponding to all phases during the event of PG fault, when current and voltage signals are recorded at PCC. It can be concluded that FI corresponding to faulty phase (phase A) has a higher magnitude compared to TM after the incidence of PG fault. However, the FI corresponding to the healthy phases B and C has a lower magnitude as compared to TM. Hence, it is established that algorithm is effective for the identification of PG fault and discrimination of the healthy and faulty phases by recording the data at PCC. It establishes the applicability of algorithm to provide protection against faults for the RE sources based hybrid grids.

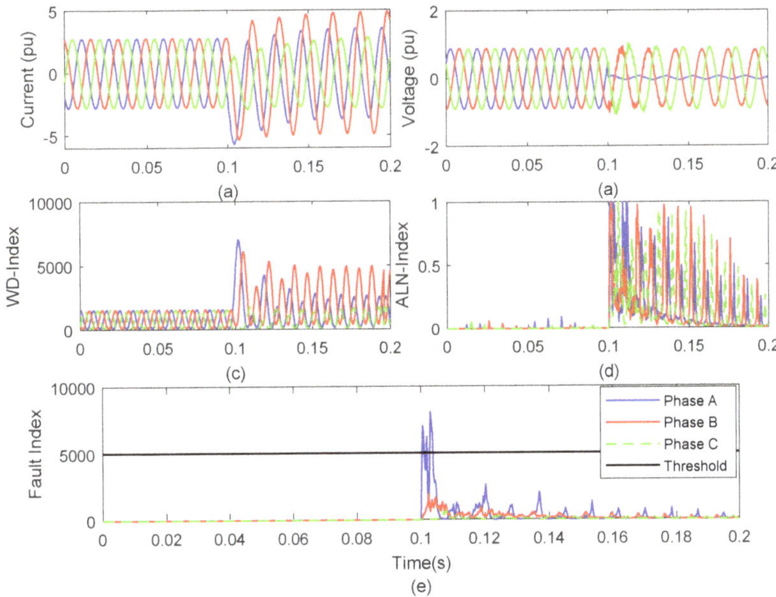

Figure 13. Recognition of the PG fault incident at node 646 by recording voltage and current signals at PCC (**a**) current waveform (**b**) voltage waveform (**c**) WD-index (**d**) ALN-index (**e**) FI.

7. Discrimination Between Switching and Faulty Events

This section details the results, obtained by the application of proposed algorithm for the switching events to differentiate these events from the faulty conditions.

7.1. Switching of Line Feeder

The event of switching on and off of the line feeder was realized by opening the circuit breaker (CB) between nodes 671 and 692 at 4th cycle and re-closing the CB at 8th cycle. Results for the period of 12 cycles in the event of switching of line feeder are presented in Figure 14. It is observed that the current and voltage waveforms are distorted with small magnitude at the time of feeder operation (on/off) as depicted in Figure 14a,b, respectively. It is inferred from Figure 14c that magnitude of WD-index increases at the time of feeder tripping and re-closing. Figure 14d indicates that the ALN-index has high magnitude peaks at the time of re-closing of the feeder. However, small magnitude transients are observed at the moment of feeder tripping. Figure 14e gives an indication that FI has a lower magnitude compared to threshold at both the events of tripping as well as re-closing of feeder. However, small magnitude peaks are observed at the moment of feeder tripping as well as re-closing, where feeder re-closing event has a large effect compared to tripping event. A high resolution plot of FI is illustrated in Figure 14f.

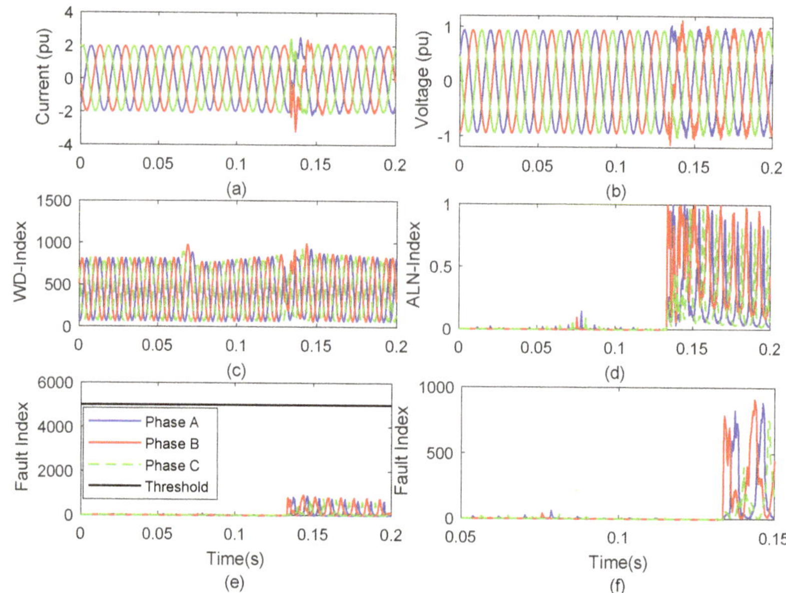

Figure 14. Recognition of event of switching of line feeder (**a**) current waveform (**b**) voltage waveform (**c**) WD-index (**d**) ALN-index (**e**) FI (**f**) high resolution plot of FI.

7.2. Capacitive Switching

An event of capacitor switching is realized at node 675 by switching off the connected capacitor banks at 4th cycle and again switching on at 8th cycle. Voltage and current signals recorded on node 650 are processed to obtain the FI using WD-index and ALN-index as illustrated in Figure 15. It is inferred from Figure 15 that FI has lower values compared to the threshold during the event of switching off and on of the capacitor bank. Hence, it is established that algorithm is effective for the discrimination of the capacitor switching event from the faulty events.

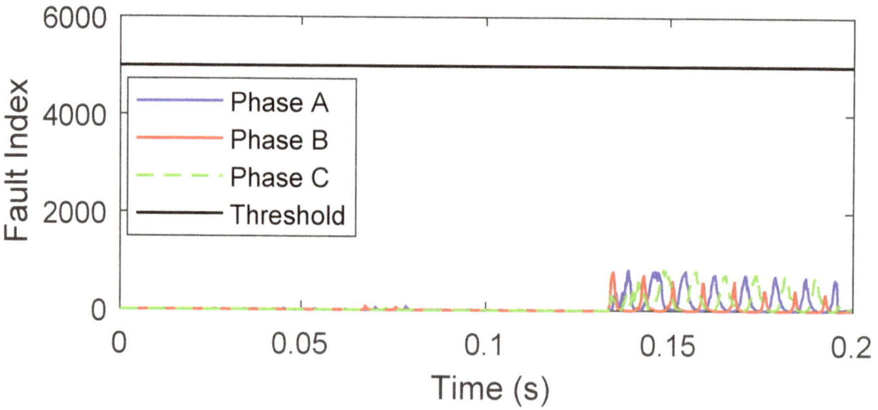

Figure 15. Fault index with event of capacitor switching.

7.3. Load Switching

An event of load switching is realized by switching off the load connected at node 671 at 4th cycle and again switched on at 8th cycle. Voltage and current signals recorded at node 650 are processed to obtain the FI using WD-index and ALN-index as illustrated in Figure 16. It is inferred from Figure 16 that FI has lower values compared to the threshold during the event of switching off and on the load. Hence, it is established that the algorithm is effective for the discrimination of load switching event from the faulty events.

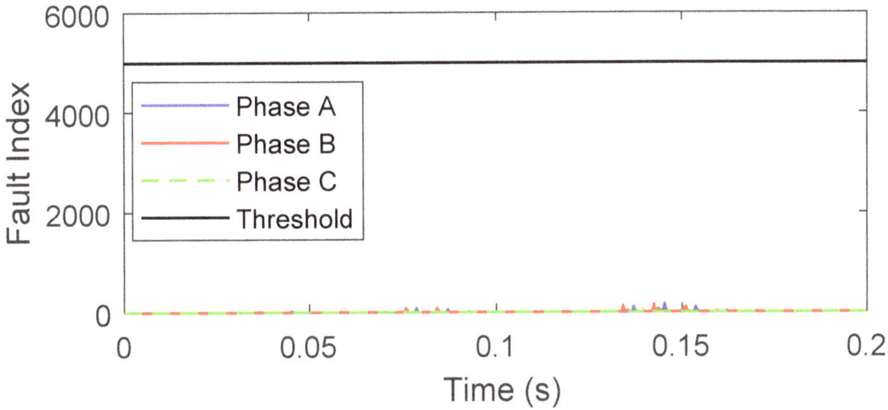

Figure 16. Fault index with event of load switching.

8. Performance of Protection Scheme

This section presents brief comments on the results and comparison of results with the existing algorithms.

8.1. Brief Discussion of Results

Results discussed in the Sections 4–7 show that the algorithm introduced in this paper is effective in providing protection against different types of faults to the hybrid grids where conventional generator, wind power plant, solar energy plant, loads, OH lines, UG cables, capacitors, transformers, measuring devices and switches are available. This is effective only if both voltage and current signals are used. It will not work when either current or voltage signals are used individually. This algorithm is fast and accurately identifies as well as classifies all types of faults during different scenarios such as presence of high level noise, different fault incidence angles, high fault impedance, different sampling frequencies, different locations of the RE generators, and different fault locations. The algorithm is effective for protection of the high voltage networks up to voltage levels of 765 kV which comprises of thermal plants, hydro plants, nuclear plants, wind generators integrated to network at different nodes, solar generators integrated to network at different nodes, feeding agriculture (induction motors) loads, domestic loads and industrial loads. However, for ring systems the protection scheme has to be installed at both ends of the line. Results also show that algorithm is effective in discriminating the faulty and operational events.

8.2. Comparative Study

Performance of the proposed algorithm is compared with the algorithms used for the detection of faults in the presence of renewable energy using DWT [12,13] and Stockwell transform (ST) [14,15]. It is observed that the DWT-based technique effectively detects the faults; however its performance is adversely affected by the presence of noise above 50 dB SNR and generates false tripping signals.

Hence, DWT-based technique is not effective for the detection of faults when high noise level of 10 dB SNR is present. The Stockwell transform-based algorithm uses the patterns of various contours and plots for the fault detection, which ranges from half cycle to one cycle. The protection scheme based on the application of proposed WD-index individually will be slow and fault detection time will be high. However, its performance is least affected by the noise. The proposed ALN-index is effective for reducing the fault detection time because it has the merit of sharp change at the time of fault incidence. However, this cannot be used as a standalone protection scheme because at the time of fault incidence, its magnitude increases for the faulty as well as healthy phases. Furthermore, the performance of ALN-index-based method is adversely affected by the presence of high noise level. Therefore, proposed algorithm, obtained using WD-index and ALN-index has the combined merits of these indexes and provide fast performance of protection scheme, further it is not affected by the noise. Hence it is established that the performance of proposed algorithm is superior as compared to DWT, Stockwell transform, standalone WD-index and standalone ALN-index.

9. Conclusions

A method for the detection and classification of faults is presented in this paper to design a protection scheme for hybrid grid system integrated with wind and solar energy plants. Various fault types such as PG, 2P, 2PG, and 3PG were successfully identified in the RE sources based hybrid grid using proposed fault index. The FI is obtained using Wigner distribution function based decomposition of current signals (WD-index) sample with alienation coefficient of voltage signals (ALN-index). Fault types were classified using number of faulty phases and identified using proposed FI. A ground fault index, based on the Wigner distribution function supported decomposition of negative sequence current is proposed to differentiate the 2P and 2PG faults. It is concluded that the algorithm is fast and accurate for detection of all the types of faults in the hybrid grid with RE penetration. Furthermore, the algorithm was tested and effectively works in the presence of high level noise, different fault incidence angles, high fault impedance and different fault locations. The algorithm is also effective for the discrimination of switching events of capacitor, load and line feeder from faulty events. The algorithm is also effective in recognizing faults in the presence of transformers and hybrid lines consisting of OH line and UG cable sections. It is also established that the proposed algorithm obtained using WD-index and ALN-index has the combined merits of these indexes and provide fast response protection scheme whose performance is not affected by the noise. Hence it is established that performance of proposed algorithm is superior compared to DWT, ST, standalone WD-index and standalone ALN-index. As a future enhancement, testing of performance of algorithm with data recorded on real time network of the utility grids may be considered before implementation in the protection schemes.

Author Contributions: Conceptualization, S.R.O., O.P.M., A.S., V.S. and S.K.G.; methodology, S.R.O. and O.P.M.; software, S.R.O. and O.P.M.; validation, S.R.O. and O.P.M.; formal analysis, S.R.O., B.K. and O.P.M.; investigation, S.R.O. and O.P.M.; resources, A.S., S.K.G., H.H.A. and P.S.; data curation, S.R.O., B.K. and O.P.M.; writing—original draft preparation, S.R.O., B.K. and O.P.M.; writing—review and editing, B.K., H.H.A. and P.S.; visualization, B.K., H.H.A. and P.S.; supervision, A.S., S.K.G., H.H.A. and P.S. All authors have read and agreed to the published version of the manuscript.

Funding: External funding is not received for this research.

Conflicts of Interest: The authors declare that they have no known conflicts of interest.

Abbreviations

Abbreviations used in this article are detailed below:

2P	Double phase
2PG	Double phase to ground
3PG	Three phase to ground
PG	Phase to ground
ALN	Alienation
ANN	Artificial neural network
CB	Circuit breaker
DFIG	Doubly fed induction generator
DG	Distributed generation
DWT	Discrete Wavelet transform
ES	Expert system
EV	Electric vehicle
FI	Fault index
FIA	Fault incidence angle
FRDFT	Fast recursive discrete Fourier transform
FST	Fuzzy set theory
GFI	Ground fault index
ICT	Interconnecting transformer
IEEE	Institute of Electrical and Electronics Engineers
OH	Overhead
MATLAB	Matrix laboratory
MPPT	Maximum power point tracking
PCC	Point of common coupling
PQ	Power quality
PS	Protection scheme
PV	Photovoltaic
RE	Renewable energy
SF	Sampling frequency
SNR	Signal to noise ratio
ST	Stockwell transform
TM	Threshold magnitude
TMGFI	Threshold magnitude of ground fault index
UG	Underground
WECS	Wind energy conversion system
WD	Wigner distribution
WDF	Wigner distribution function

References

1. Aftab, M.A.; Hussain, S.S.; Ali, I.; Ustun, T.S. Dynamic protection of power systems with high penetration of renewables: A review of the traveling wave based fault location techniques. *Int. J. Electr. Power Energy Syst.* **2020**, *114*, 105410. doi:10.1016/j.ijepes.2019.105410. [CrossRef]
2. Eissa, M.; Awadalla, M.H. Centralized protection scheme for smart grid integrated with multiple renewable resources using Internet of Energy. *Glob. Trans.* **2019**, *1*, 50–60. doi:10.1016/j.glt.2019.01.002. [CrossRef]
3. Fang, Y.; Jia, K.; Yang, Z.; Li, Y.; Bi, T. Impact of Inverter-Interfaced Renewable Energy Generators on Distance Protection and an Improved Scheme. *IEEE Trans. Ind. Electron.* **2019**, *66*, 7078–7088. doi:10.1109/TIE.2018.2873521. [CrossRef]
4. Barra, P.; Coury, D.; Fernandes, R. A survey on adaptive protection of microgrids and distribution systems with distributed generators. *Renew. Sustain. Energy Rev.* **2020**, *118*, 109524. doi:10.1016/j.rser.2019.109524. [CrossRef]
5. Eissa, M. Protection techniques with renewable resources and smart grids—A survey. *Renew. Sustain. Energy Rev.* **2015**, *52*, 1645–1667. doi:10.1016/j.rser.2015.08.031. [CrossRef]

6. Alhelou, H.H.; Golshan, M.H.; Askari-Marnani, J. Robust sensor fault detection and isolation scheme for interconnected smart power systems in presence of RER and EVs using unknown input observer. *Int. J. Electr. Power Energy Syst.* **2018**, *99*, 682–694. doi:10.1016/j.ijepes.2018.02.013. [CrossRef]

7. Kumar, D.S.; Srinivasan, D.; Reindl, T. A Fast and Scalable Protection Scheme for Distribution Networks With Distributed Generation. *IEEE Trans. Power Deliv.* **2016**, *31*, 67–75. doi:10.1109/TPWRD.2015.2464107. [CrossRef]

8. Pavlatos, C.; Vita, V.; Ekonomou, L. Syntactic pattern recognition of power system signals. In Proceedings of the 19th WSEAS International Conference on Systems (Part of CSCC'15), Zakynthos Island, Greece, 16–20 July 2015; pp. 16–20.

9. Pavlatos, C.; Vita, V. Linguistic representation of power system signals. In *Electricity Distribution*; Springer: Berlin, Germany, 2016; pp. 285–295.

10. Nieto, A.; Vita, V.; Maris, T.I. Power quality improvement in power grids with the integration of energy storage systems. *Int. J. Eng. Res. Technol.* **2016**, *5*, 438–443.

11. Sheesh, R.O.; Amit, S.; Sunil, K.G.; Jhajharia, S.K.; Bhuvnesh, R.; Om Prakash, M. Wigner Distribution Function and Alienation Coefficient Based Transmission Line Protection Scheme. *IET Gener. Transm. Distrib.* **2020**, *14*, 1–12.

12. Suman, T.; Mahela, O.P.; Ola, S.R. Detection of transmission line faults in the presence of wind power generation using discrete wavelet transform. In Proceeidngs of the 2016 IEEE 7th Power India International Conference (PIICON), Bikaner, India, 25–27 November 2016; pp. 1–6. doi:10.1109/POWERI.2016.8077174. [CrossRef]

13. Suman, T.; Mahela, O.P.; Ola, S.R. Detection of transmission line faults in the presence of solar PV generation using discrete wavelet. In Proceeidngs of the 2016 IEEE 7th Power India International Conference (PIICON), Bikaner, India, 25–27 November 2016; pp. 1–6. doi:10.1109/POWERI.2016.8077203. [CrossRef]

14. Thukral, S.; Mahela, O.P.; Kumar, B. Detection of transmission line faults in the presence of solar PV system using stockwell's transform. In Proceedings of the 2016 2016 IEEE 7th Power India International Conference (PIICON), Bikaner, India, 25–27 November 2016; pp. 1–6. doi:10.1109/POWERI.2016.8077304. [CrossRef]

15. Ola, S.; Saraswat, A.; Goyal, S.; Jhajharia, S.; Mahela, O.P. Detection and Analysis of Power System Faults in the Presence of Wind Power Generation Using Stockwell Transform Based Median. In *Intelligent Computing Techniques for Smart Energy Systems*; Lecture Notes in Electrical Engineering; Kalam A., Niazi K., Soni A., Siddiqui S., Mundra A., Eds.; Springer: Singapore, 2020; Volume 607, pp. 319–330. doi:10.1007/978-981-15-0214-9_36. [CrossRef]

16. Kersting, W.H. Radial distribution test feeders. *Power Syst. IEEE Trans.* **1991**, *6*, 975–985. doi:10.1109/59.119237. [CrossRef]

17. Kersting, W. Radial distribution test feeders. In Proceedings of the Power Engineering Society Winter Meeting, Columbus, OH, USA, 1 February 2001; Volume 2, pp. 908–912. doi:10.1109/PESW.2001.916993. [CrossRef]

18. Mahela, O.P.; Shaik, A.G. Power quality improvement in distribution network using {DSTATCOM} with battery energy storage system. *Int. J. Electr. Power Energy Syst.* **2016**, *83*, 229–240. doi:10.1016/j.ijepes.2016.04.011. [CrossRef]

19. Paz, M.C.R.; Ferraz, R.G.; Bretas, A.S.; Leborgne, R.C. System unbalance and fault impedance effect on faulted distribution networks. *Comput. Math. Appl.* **2010**, *60*, 1105–1114. doi:10.1016/j.camwa.2010.03.067. [CrossRef]

20. Shaik, A.G.; Mahela, O.P. Power quality assessment and event detection in hybrid power system. *Electric Power Syst. Res.* **2018**, *161*, 26–44. doi:10.1016/j-eps-converted-to.pdfr.2018.03.026. [CrossRef]

21. Mahela, O.P.; Shaik, A.G. Power Quality Detection in Distribution System with Wind Energy Penetration Using Discrete Wavelet Transform. In Proceedings of the 2015 Second International Conference on Advances in Computing and Communication Engineering (ICACCE), Dehradun, India, 1–2 May 2015; pp. 328–333. doi:10.1109/ICACCE.2015.52. [CrossRef]

22. Mahela, O.P.; Khan, B.; Alhelou, H.H.; Siano, P. Power Quality Assessment and Event Detection in Distribution Network with Wind Energy Penetration Using Stockwell Transform and Fuzzy Clustering. *IEEE Trans. Ind. Inform.* **2020**, 1–10, In press. [CrossRef]

23. Kuo, C.L.; Lin, C.H.; Yau, H.T.; Chen, J.L. Using Self-Synchronization Error Dynamics Formulation Based Controller for Maximum Photovoltaic Power Tracking in Micro-Grid Systems. *Emerg. Sel. Top. Circ. Syst. IEEE J.* **2013**, *3*, 459–467. doi:10.1109/JETCAS.2013.2272839. [CrossRef]
24. Ding, K.; Bian, X.; Liu, H.; Peng, T. A MATLAB-Simulink-Based PV Module Model and Its Application Under Conditions of Nonuniform Irradiance. *Energy Convers. IEEE Trans.* **2012**, *27*, 864–872. doi:10.1109/TEC.2012.2216529. [CrossRef]
25. Mahela, O.P.; Shaik, A.G. Power quality recognition in distribution system with solar energy penetration using S-transform and Fuzzy C-means clustering. *Renew. Energy* **2017**, *106*, 37–51. doi:10.1016/j.renene.2016.12.098. [CrossRef]
26. Khan, N.A.; Taj, I.A.; Jaffri, M.N.; Ijaz, S. Cross-term elimination in Wigner distribution based on 2D signal processing techniques. *Signal Process.* **2011**, *91*, 590–599. doi:10.1016/j.sigpro.2010.06.004. [CrossRef]
27. Cheng, J.Y.; Huang, S.J.; Hsieh, C.T. Application of Gabor–Wigner transform to inspect high-impedance fault-generated signals. *Int. J. Electr. Power Energy Syst.* **2015**, *73*, 192–199. doi:10.1016/j.ijepes.2015.05.010. [CrossRef]
28. Rathore, B. Alienation based fault detection and classification in transmission lines. In Proceedings of the 2015 Annual IEEE India Conference (INDICON), New Delhi, India, 17–20 December 2015; pp. 1–6. doi:10.1109/INDICON.2015.7443649. [CrossRef]
29. Rathore, B.; Shaik, A.G. Wavelet-alienation based protection scheme for multi-terminal transmission line. *Electric Power Syst. Res.* **2018**, *161*, 8–16. doi:10.1016/j-eps-converted-to.pdfr.2018.03.025. [CrossRef]

Article

Addressing Abrupt PV Disturbances, and Mitigating Net Load Profile's Ramp and Peak Demands, Using Distributed Storage Devices

Roshan Sharma* and Masoud Karimi-Ghartemani

Department of Electrical and Computer Engineering, Mississippi State University, Starkville, MS 39762, USA; karimi@ece.msstate.edu

* Correspondence: rs2142@msstate.edu

Received:10 January 2020; Accepted: 21 February 2020; Published: 25 February 2020

Abstract: At high penetration level of photovoltaic (PV) generators, their abrupt disturbances (caused by moving clouds) cause voltage and frequency perturbations and increase system losses. Meanwhile, the daily irradiation profile increases the slope in the net-load profile, for example, California duck curve, which imposes the challenge of quickly bringing on-line conventional generators in the early evening hours. Accordingly, this paper presents an approach to achieve two objectives: (1) address abrupt disturbances caused by PV generators, and (2) shape the net load profile. The approach is based on employing battery energy storage (BES) systems coupled with PV generators and equipped with proper controls. The proposed BES addresses these two issues by realizing flexible power ramp-up and ramp-down rates by the combined PV and BES. This paper presents the principles, modeling and control design aspects of the proposed system. A hybrid dc/ac study system is simulated and the effectiveness of the proposed BES in reducing the impacts of disturbances on both the dc and ac subsystems is verified. It is then shown that the proposed PV-BES modifies the daily load profile to mitigate the required challenge for quickly bringing on-line synchronous generators.

Keywords: PV generator; power ramping; battery energy storage; duck curve; hybrid dc/ac grids

---

1. Introduction

The limitations of fossil fuel reserves and environmental concerns have increased interest in renewable energy resources [1]. Solar photovoltaic (PV) generation is one of the most promising renewable energy approaches. In recent years, PV power generation has increased significantly thanks to the development of cost-effective, higher efficiency semiconductor PV technology, as well as advancements in related power electronics and controls. Solar PV is projected to contribute 16% of the world's total electricity and 43% of the added electricity in the United States (US) by 2050 [2,3]. However, the uncertainty and variability of PV sources constitute the main obstacles against their high level penetration. Furthermore, PV sources do not have inherent rotational inertia to support the grid during the transients [4–6].

The power injected by PV can change suddenly due to cloud movements. In 2011, a study of three PV systems, located in Porterville, Palmdale, and Fontana, California, found that the power ramp rate was as high as 90%, 93% and 75% per minute of their rated power, respectively [7]. The sudden changes in PV power can introduce voltage fluctuations in distribution system and frequency fluctuations in the case of high PV penetration and weak grid, leading to grid stability issues [2,8–10]. Puerto Rico Electric Power Authority (PREPA) limits the ramp rate to 10%/minute of the rated capacity for both wind and PV generation, and the Hawaiian Electric Company (HECO) limits its ramps at ±2 MW/min (even ±1 MW/min during some times) for wind generators under 50 MW [11]. According to IEEE standard 1547, distributed energy resources (DERs) cannot cause voltage fluctuations exceeding 3% for a medium voltage interconnection and 5% for a low voltage interconnection [12].

A large number of distributed PV generators together create a duck-shaped daily net load profile due to the variation in solar irradiation. The duck-shaped net load profile has steep slopes, especially in the evening. This requires bringing a large number of synchronous resources online within a short period of time. Such changes in the net load profile cannot be easily addressed by slow conventional generators and the problem becomes severe with the increased penetration of PVs. The existence of a large number of PV generators also increases the chances of negative net load that requires curtailment of PV power [8,13,14].

The number of PV generators connected to the electric power grid has been rising continuously. Authorities and researchers are considering the development of strategies and approaches to obtain grid services, both transient and long-term, from the aggregation of these distributed resources [15]. Therefore, PV generators have been coupled together with complementary generation and storage technologies to address the intermittencies and the disturbances, and these distributed resources are utilized to obtain grid services. Battery energy storage (BES) systems and super-capacitors are two important energy storage approaches that can be coupled with PV generators [16–18].

Most of the existing approaches that combine BES with household PV-prosumers are concerned with optimization of battery size, and energy management strategies to minimize electricity bills [19–21]. These methods do not address the impact of intermittency and power variation of PVs on the grid. Moving average methods are used to mitigate the impact of abrupt PV power disturbances. However, moving average method does not directly control the ramp rate, depends on historical data, and operates the BES even though BES operation is not required due to the memory effect [22]. The direct ramp control of PV power is an effective method. The ramp-up of PV output is generally not a challenging issue, and it can be achieved using a proper controller without requiring a storage, at the expense of total extracted PV energy [23]. However, the ramp-down is a challenge, and it requires an energy storage to compensate for the PV power drop.

The fluctuations of wind/PV generators are compensated using the BES station in Reference [24]. The control approach calculates the target power for the BES station from the fluctuations of wind/PV generators, and this power is shared by the constituting smaller BES units based on their SOC level. In Reference [22], a BES is coupled with a PV system to mitigate the power fluctuations caused by cloud passing using the ramping method. During PV power fluctuations, the BES is controlled based on the inverse characteristics of the desired ramp-rate. Once the PV power fluctuation is over, the BES power is brought to zero using SOC droop-based ramp rate. The control strategies of References [22,24] are presented at system level, and ignore the dynamics of the converter and details of the control design. Furthermore, the BES reference calculations involve complex algorithms. A hybrid BES and super-capacitor setup is used in Reference [25] to provide frequency support. Although this is combined with a PV generator, it does not directly address the PV power generation variations. Moreover, the frequency does not directly and quickly reflect the PV disturbances at the distribution level.

A battery is integrated in an ac-stack architecture in series with PV converters in Reference [26] to achieve the controlled ramp rate in power. The BES control uses the derivative of the rms of ac string current to determine the fluctuation in PV power. This method does not appear to be able to respond to abrupt PV fluctuations in a timely manner due to the limitations with a derivative operator in practice. Furthermore, this architecture is vulnerable to over-modulation that distorts the output. Finally, the stack architecture compromises the plug-and-play flexibility of the system. In Reference [27], a power management strategy is proposed for a BES in connection to a PV and a local load. If the local load is larger than a predefined minimum value, the combined PV-BES supply the load and the rest of power is offset by the grid. No power ramping is done in this mode. If the local load is less than a predefined minimum, the battery performs a ramp rate control, based on Reference [22], if the PV fluctuations exceed the predefined ramp rate; and the PV-BES operates in constant power feed mode otherwise. Further details of the control design procedures are not presented. A PV-battery combination connected to an ac grid is presented in Reference [28] where the PV operates at the MPP, a battery controls the dc bus voltage of the inverter and the inverter operates in constant power mode. If the reference power of the inverter is properly determined, it can realize ramping. However, this is not discussed in Reference [28]. One limitation of this approach is that the BES cannot be used for an already operating inverter without altering the control system of the inverter. A similar approach is used in Reference [29] where the BES supplies/absorbs the difference of power between the desired and the PV power. The controller in Reference [29] requires communication of an ac signal which is a major limitation. In Reference [30], a 1 MW/2 MWh BES is connected with a multi-MW PV system to achieve (1) power smoothing using moving average and low pass filter, and (2) frequency regulation using P-f droop control. The power smoothing method of Reference [30] has the disadvantages of the moving average approach mentioned earlier. Also, the frequency regulation may not be useful at the distribution level.

Most of the aforementioned works that implement BES to smooth PV power treat the matter at the system level and ignore the dynamics at the converters level. There is a lack of systematic design of the local controls to accomplish the desired power ramps. This paper presents a BES control system concept to address both the abrupt and daily variations of PV power. The proposed BES connects at the terminals of the PV generator (without requiring a change in the PV inverter) and responds to the disturbances to execute desired power ramp rates. The ramp rate signals are either generated locally or received from the grid operator. The paper presents the complete mathematical model of the system including the dynamics of the converter and provides systematic approach in designing the control. Furthermore, if the BES is only used to filter the abrupt PV disturbances for a short period, it results in low utilization of the BES. Therefore, the ramping capability of the PV-BES system has been utilized in this paper to improve the daily load profile. The grid operator can remotely command the distributed PV-BES units with the desired ramp rate, and the proposed control executes the command. The aggregated effect of the distributed PV-BES units with the proposed ramping capability significantly enhances the daily net load profile in terms of (1) reduced peak, (2) increased minimum, and (3) reduced slopes. The proposed method also includes an outer loop that controls the SOC of the BES. The performance of the proposed system is verified using extensive computer simulation and a laboratory prototype.

The contributions of this paper may be summarized as follows: (1) A simple yet effective battery control philosophy to address both the abrupt and daily photovoltaic generation variations. The two set-points of the proposed controller can be generated locally or through communications for system-wide optimization; (2) Complete modeling and control design of the battery control system; (3) Verification of the proposed approach using simulations and laboratory tests.

The remainder of the paper is organized as follows. The power variabilities, abrupt and daily profile, originated from PV generators are discussed in Section 2. Section 3 presents a proposed battery energy storage (BES) system, its mathematical model, and the proposed control to ramp the output power of the PV. Numerical designs and the results obtained both in simulation and laboratory experiment are presented in Section 4. The paper concludes with concluding remarks and limitations of the proposed control in Section 5.

2. Power Variabilities of a PV Generator

2.1. Abrupt Disturbances in PV Generation

Cloud movements cause abrupt changes in solar irradiation, Figure 1. To study the impact of these abrupt power changes, a hybrid dc/ac system is considered in this paper as shown in Figure 2. The parameters of the dc microgrid and the ac system are shown in Tables 1 and 2, respectively. The details of the PV control used in this study are given in Section 4.1.1.

The PV generator of Figure 2 is controlled using a maximum power point tracking (MPPT) algorithm. The control structure and the parameters are shown in Section 4. Figure 3 shows the voltage and frequency fluctuations, at the point of coupling between dc and ac systems, to abrupt PV disturbances. The solar intensity falls from 1000 W/m^2 to 250 W/m^2 at $t = 5.0$ s, and it rises back to 1000 W/m^2 at $t = 50.0$ s. The dc voltage has a peak of 10% that oscillates over a second, and the ac voltage (rms) has a peak close to 5%. The frequency oscillates with a peak of 0.23 Hz, and it takes several seconds to settle.

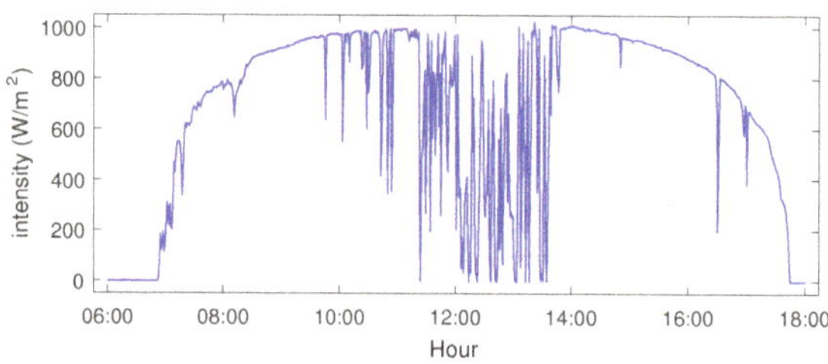

Figure 1. Solar intensity measured at Lowry Range Solar Station on 18 March 2013 [31].

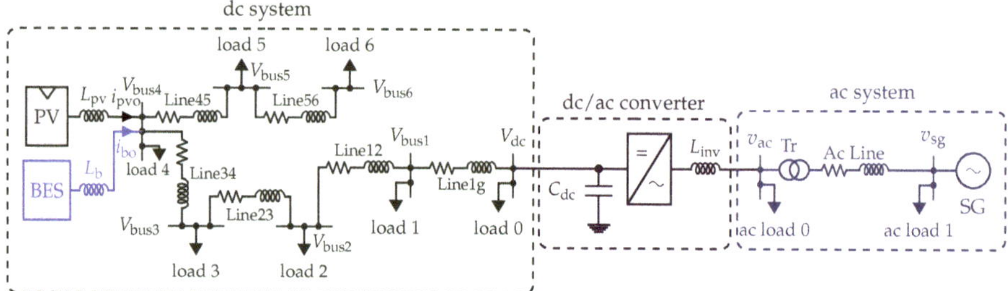

Figure 2. A hybrid dc/ac system to study the abrupt power disturbances of the photovoltaic (PV) generator.

Table 1. Parameters of dc system.

Nominal bus Voltage: 600 V			
Loads:			
Load 0: 1.5 kW	Load 1: 0.5 kW	Load 2: 1.5 kW	
Load 3: 2.0 kW	Load 4: 1.5 kW	Load 5: 5.0 kW	
Load 6: 4.4 kW			
Lines:			
Line	Length	Resistance	Inductance
Line1g	0.75 km	0.482 Ω	0.198 mH
Line12	0.5 km	0.321 Ω	0.132 mH
Line23	0.3 km	0.0.193 Ω	0.079 mH
Line34	0.5 km	0.321 Ω	0.132 mH
Line45	0.5 km	0.321 Ω	0.132 mH
Line56	1 km	0.642 Ω	0.264 mH
PV system:			
Rated power: 15 kW at 1000 W/m^2, 1000 V			

Table 2. Parameters of ac system.

Synchronous Generator (SG):

Power (S_{sg}): 250 kVA, voltage (v_{sg}): 12.47 V
Inertia constant (H): 5.0

Loads:
ac load 0: 0.05 pu,   ac load 1: 0.846 pu

Ac Line:
200 km, 32.2 Ω, 120.96 mH

Transformer:
Resistance (R_{tr}): 0.02 pu, Reactance (X_{tr}): 0.05 pu
Voltage ratio (VR): 207.8 V / 12.47 kV

Three-phase dc/ac Converter:

Parameter	Symbol	Value
Filter inductance	L_{inv}	10 mH
Filter resistance	R_{inv}	10 mΩ
Dc capacitor	C_{dc}	1200 µF
Voltage on dc side	V_{dc}	600 V
Voltage on ac side	V_{ac}	207.8 V

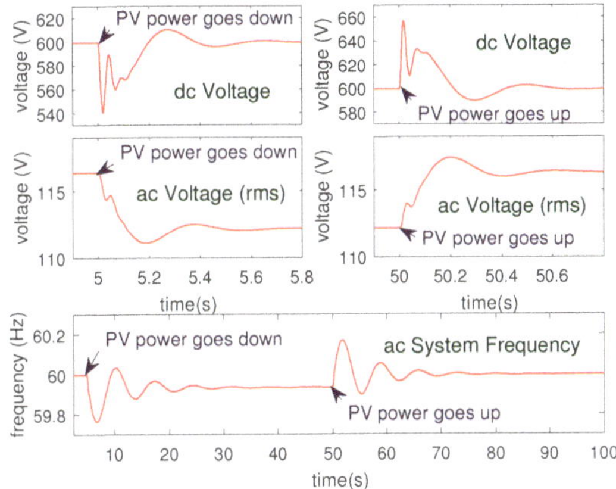

Figure 3. Voltage and frequency fluctuations at point of common coupling (PCC) of the study system due to the abrupt PV disturbances.

2.2. Daily Power Profile of PV

Figure 4a shows the power generation and the load demand of California Independent System Operator (ISO) on 31 March 2019. As the sun rises, the PV generation increases and the net load declines. Into the evening, the PV generation declines while the load demand increases. This creates a duck-shaped net load profile with steep ramp rates. The net profile of Figure 4a has a ramp rate as steep as 5300 MW/h. Such power ramps impose challenges in commitment of the slow conventional generators [32]. The problem becomes severe with the increased penetration level of the PVs as shown in Figure 4b.

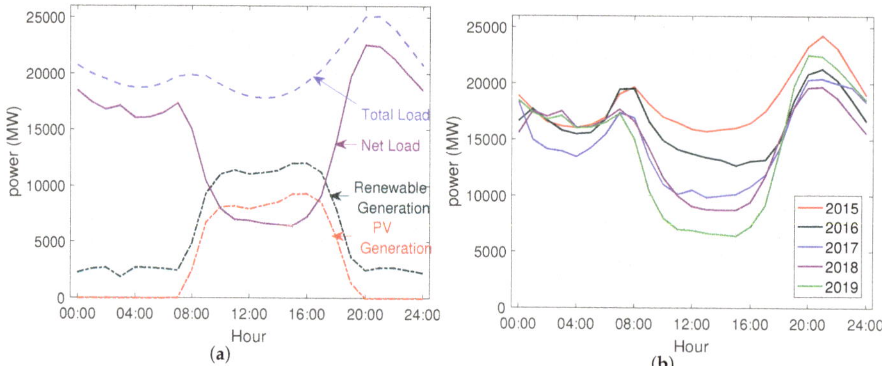

Figure 4. Load profile of California ISO; (**a**) Load demand and generation on 31 March 2019, and (**b**) net load profile on March 31 for last 5 years (2015-19) [33].

3. Proposed BES System and Control

Figure 5 shows the structure of the proposed PV-BES system. The BES connects at the terminals of the PV and executes ramping of the combined power of the PV-BES system. The BES either generates locally or receives the ramp-up-rate (RUR) and ramp-down-rate (RDR) signals from the grid operator and adjusts its operation accordingly. This can (1) prevent the abrupt PV disturbances, and (2) re-shape the daily load profile, as will be demonstrated in this paper.

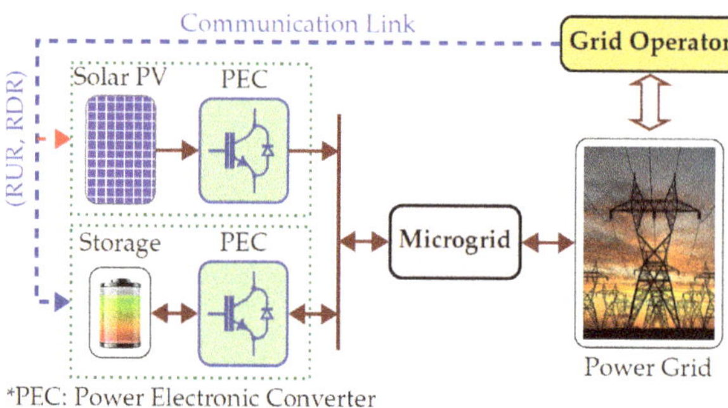

Figure 5. Structure of proposed PV-battery energy storage (BES).

3.1. Defining Operational Scenario of BES

Figure 6a shows the scenario where PV power (P_{pvo}) experiences an abrupt fall followed by an abrupt rise. The BES power, shown in Figure 6b, compensates this sudden disturbance and the combined power of PV-BES system, P_o, has controlled ramp down/up. The charging and discharging of the BES causes rise/fall of its state-of-charge (SOC), as shown in Figure 6c. The BES control system should be equipped with an SOC control that charges/discharges the battery after the response to the disturbance is completed to regulate its SOC to the desired level.

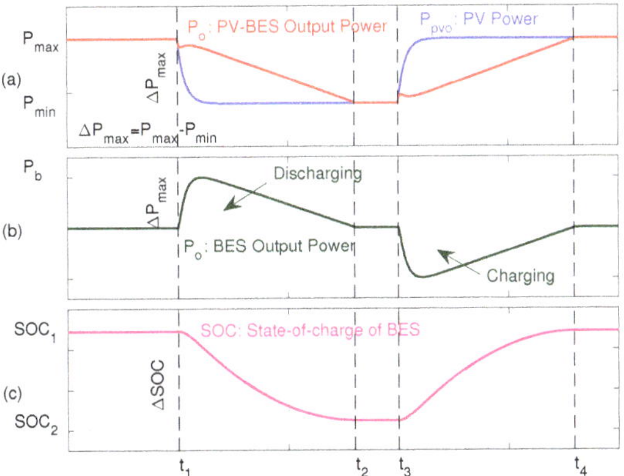

Figure 6. Operational scenario of the BES.

3.2. BES Converter and Its Mathematical Modeling

In this study, the BES uses a bi-directional converter topology as shown in Figure 7. The control system determines the duty cycle of the gating signal for the switches. Assume $d_b(t)$ is the duty cycle of PWM of the BES converter and $v_b(t)$ is the BES voltage, then the average converter voltage is $v_{bo}(t) = d_b(t)v_b(t)$. If $u_b(t) = v_{bo}(t)$ is the control input, then $d_b(t) = \frac{u_b(t)}{v_b(t)}$ and

$$\frac{di_{bo}(t)}{dt} = -\frac{R_b}{L_b}i_{bo}(t) + \frac{1}{L_b}u_b(t) - \frac{V_{bus}}{L_b}, \tag{1}$$

where $i_{bo}(t)$, R_b, L_b, and V_{bus} are the BES converter output current, filter resistance, filter inductance, and the voltage at the point of connection of the BES converter, respectively. In terms of power, (1) can be rewritten as

$$\frac{dP_{bo}(t)}{dt} = -\frac{R_b}{L_b}P_{bo}(t) + \frac{V_{bus}}{L_b}u_b(t) - \frac{V_{bus}^2}{L_b}, \tag{2}$$

where $P_{bo}(t)$ is the output power from the BES system. The dynamics of the SOC depends on the amount of current flowing in/out of the battery and it is mathematically expressed as

$$\frac{d}{dt}\text{SOC}(t) = -\frac{P_{bo}}{Qv_b}, \tag{3}$$

where Q and v_b are the battery capacity (in As) and battery voltage (in V), respectively.

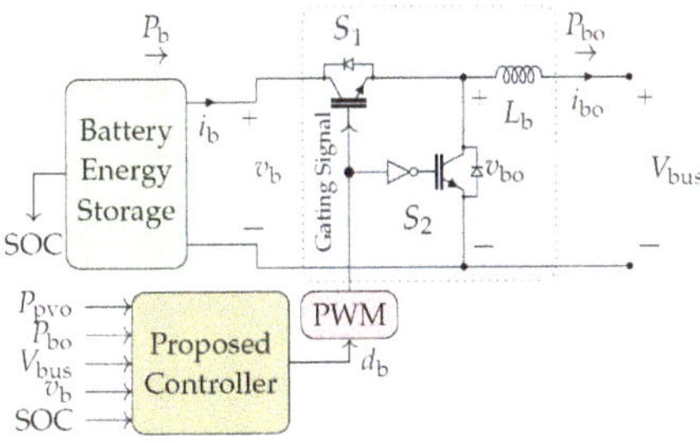

Figure 7. BES converter system.

3.3. BES Control Structure and Design

Figure 8 shows the block diagram of the proposed control structure for the BES system. The control system can be divided into 3 parts: the Power Control (high bandwidth), the Reference Power Calculator, and the SOC control (low bandwidth). These three parts of the controller can be designed separately and independently due to their different time scales.

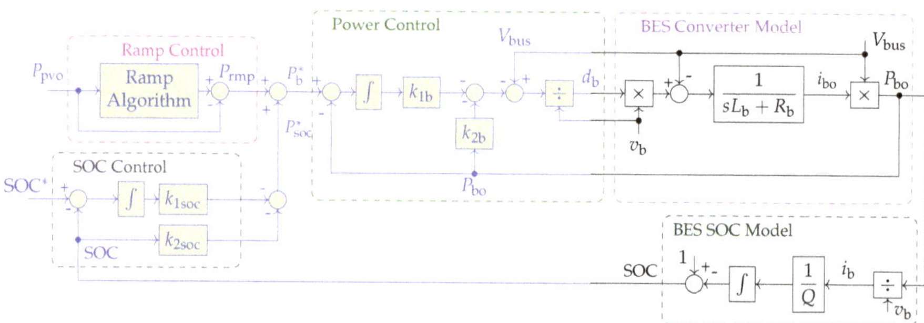

Figure 8. Mathematical model of the BES converter system with the proposed controller.

3.3.1. Power Control Loop

This loop generates the duty cycle for the converter to absorb/supply the reference power (P_b^*). A proportional-integrating (PI) controller is considered and its gains, k_{1b} and k_{2b}, are optimally designed as follows. Define $\dot{x}_1(t) = e(t) = P_b^*(t) - P_{bo}(t)$, and $x_2(t) = P_{bo}(t)$ to obtain the state space equations of the power control loop as

$$\dot{x}_1(t) = -x_2(t) + P_b^*(t)$$
$$\dot{x}_2(t) = -\frac{R_b}{L_b}x_2(t) + \frac{V_{bus}}{L_b}u_b(t) - \frac{V_{bus}^2}{L_b}. \tag{4}$$

We use the approach [34] to convert this tracking problem into standard linear quadratic regulator (LQR) problem by applying $\frac{d}{dt}$ to both sides of (4) to obtain

$$\dot{z}_1(t) = -z_2(t),$$
$$\dot{z}_2(t) = -\frac{R_b}{L_b}z_2(t) + \frac{V_{bus}}{L_b}W_b(t), \tag{5}$$

where $z_i(t) = \dot{x}_i(t)$ and $W_b(t) = \dot{u}_b(t)$. It is assumed that $\frac{d}{dt}V_{bus}$ and $\frac{d}{dt}P_b^*$ are 0, which is justifiable. Therefore, $\dot{z}(t) = Az(t) + BW_b(t)$ and the objective is to regulate $z_1(t) = e(t)$ to 0. The cost function is

$$J = \int_0^\infty [q_1 e^2(t) + q_2 z_2^2(t) + W_b^2(t)]dt. \tag{6}$$

As described in Reference [34], the q_i parameters can be systematically adjusted for the design of the gains (k_{1b}, k_{2b}) to achieve the fast and smooth response in the power control. The numerical design stage for this system is described in Section 4.

3.3.2. Reference Power Calculation

The reference power is calculated using an algorithm to properly follow the desired ramp rate. Figure 9 shows the flowchart that generates a reference power value, P_{rmp}, that follows desired ramp rate whenever the PV power experiences changes. If P^*_{soc} is power for the SOC control, then the reference power for the BES system is

$$P^*_b(t) = P_{\text{rmp}}(t) + P^*_{\text{soc}}(t).\tag{7}$$

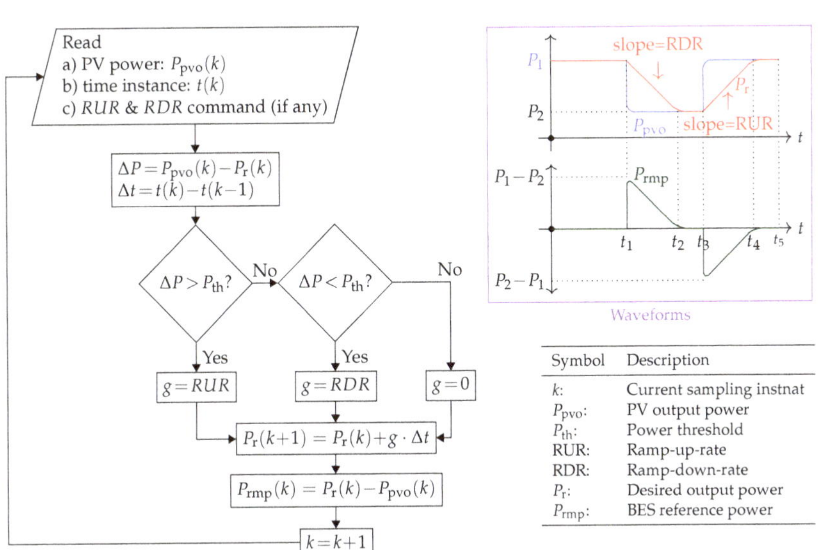

Figure 9. Flowchart to calculate power ramp reference for the BES.

3.3.3. SOC Control Loop

A PI controller is used and its gains, $k_{1\text{soc}}$ and $k_{2\text{soc}}$, are optimally designed to keep the SOC close to its reference value as shown in Figure 8 [6]. The SOC controller gains are designed such that the power control loop and the ramp reference calculator are faster than it. Therefore, during the design of this loop, the fast dynamics of the power control loop may be neglected (and this loop is substituted with a unity gain and the ramp power P_{rmp} is set to zero). With these assumption and also neglecting the small voltage variations across the battery that is, $v_b \approx V^*_b$, the system becomes linear. Defining $\dot{x}_1(t) = e(t) = \text{SOC}^* - \text{SOC}(t)$, $x_2(t) = \text{SOC}(t)$, and $u(t) = P_{\text{soc}}(t)$, the state equations of SOC control loop are

$$\dot{x}_1(t) = -x_2(t) + \text{SOC}^*,$$
$$\dot{x}_2(t) = -\frac{1}{QV^*_b}u(t).\tag{8}$$

This can be converted to a LQR form by applying $\frac{d}{dt}$ as

$$\dot{z}_1(t) = -z_2(t),$$
$$\dot{z}_2(t) = -\frac{1}{QV^*_b}W_{b1}(t)\tag{9}$$

where $z_i(t) = \dot{x}_i(t)$. Thus, $\dot{z}(t) = Az(t) + BW_{b1}(t)$ and the objective is to regulate $z_1(t) = e(t)$ to 0. Define the cost function

$$J = \int_0^\infty [q_1 e^2(t) + q_2 z_2^2(t) + W_{b1}^2(t)]dt,\tag{10}$$

and use the method of Reference [34] for the design of the gains (k_{1soc}, k_{2soc}) to achieve smooth response in the SOC control. Numerical design is provided in Section 4.

3.4. Determining the BES Capacity

The battery capacity depends on the desired slowest power ramp of the PV and the BES combined. The minimum BES capacity (Q_{min}) is chosen such that during the maximum power disturbance (ΔP_{max}) the system operates at desired slowest power ramp (R_{min}) keeping the SOC within the desired range (SOC$_{min}$, SOC$_{max}$) for all the time.

Figure 6 shows the typical desired response of the system. For maximum PV power fluctuation ΔP_{max} (in W), the BES should be able to supply power with minimum ramp R_{min} (in W/s) for the time interval T where $T = \frac{\Delta P_{max}}{R_{min}}$. Assuming that the battery voltage remains relatively constant at V_b^* (in V) during this period, the total charge supplied by the battery is $\frac{0.5 T \Delta P_{max}}{V_b^*}$ (in As). If ΔSOC_{max} is the maximum permissible fluctuation in SOC from its nominal value, SOC$_n$, during this period, then the minimum battery capacity required, Q_{min} (in As), is given by

$$Q_{min} = \frac{\Delta P_{max}^2}{2V_b^* R_{min} \Delta SOC_{max}}. \tag{11}$$

4. Numerical Designs, Results, and Comparisons

This section presents some numerical designs of the proposed system and also investigates its performance in the context of both abrupt PV power variations and daily irradiation profile using simulations and laboratory-scale experimentation. The structure of the this section is as follows. (1) Section 4.1 studies the performance of proposed system in response to abrupt PV disturbances in a simulated hybrid dc/ac system. (2) Section 4.2 shows the simulation results of applying the proposed system to address abrupt PV disturbances in a real irradiation profile data. (3) Section 4.3 investigates the performance of the proposed method in mitigating the duck-curve phenomenon using simulations on practical data. (4) Section 4.4 shows the results of a laboratory-scale realization of the proposed method. (5) Section 4.5 shows the qualitative comparison of the proposed method with existing methods to mitigate PV power fluctuations.

4.1. Abrupt Disturbances: Case Study 1

The proposed controller is applied to the study system of Figure 2 to study the impact of abrupt PV disturbances. The design of system components are discussed first.

4.1.1. System and Control Parameters

Figure 10 and Table 3 show the PV system and its converter parameters. The controller gains of the PV system are designed using the method discussed in Reference [34]. The internal current loop is designed first, which gives k_{1pv} and k_{2pv}, and then the external loop is designed that includes the design of k_{3pv}, k_{4pv}, k_{5pv}, and k_{6pv}. In the design of the external loop, all the dynamics of the internal loop are carried along. Therefore, unlike common practice, the internal loop is not limited to be several times faster than the external loop. In our study, the PV operates at MPPT using perturb and observe (P&O) algorithm [35]. However, this is not a requirement for the proposed BES controller.

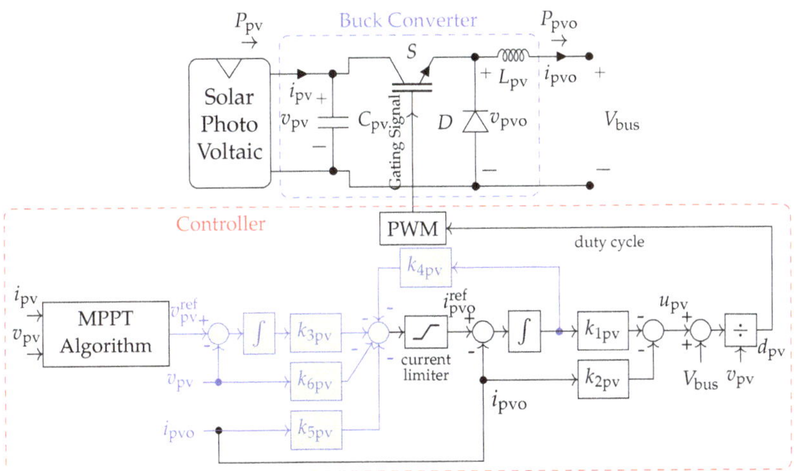

Figure 10. PV converter system and its control.

Table 3. Parameters of PV system.

PV System:
15 kW @ 1000 W/m^2 and 1000 V
V_{bus}: 600 V
Converter:
C_{pv} : 400 µF, L_{pv} : 10 mH, R_{pv}: 10 mΩ
Controller:
Switching Frequency: 10 kHz
Control Gains:
k_{1pv}: −316.22, k_{2pv}: 3.07, k_{3pv}: 31.62
k_{4pv}: 344.49, k_{5pv}: 1.38, k_{6pv}: −0.59

The objective is to obtain the desired ramp rate of 30 %/minute $\approx$ 75 W/s from the combined PV-BES system. Table 4 shows the parameters of the BES system and its converter. The gains of the power control loop of the BES controller are designed based on the method discussed in Section 3.3.1. Figure 11a shows the plot of the closed loop poles of the power loop on changing q_i's. Here, q_1 and q_2 are subsequently increased between $10^{-3} \rightarrow 10^{2.5}$ and $10^{-6} \rightarrow 10^{-4}$, respectively. The reference power of the BES for the ramp control is calculated by using the method mentioned in Section 3.3.2. For the SOC control loop, the control gains are designed based on the method presented in Section 3.3.3. Figure 11b shows the plot of the closed loop poles of the SOC loop on changing q_i's. Here, q_1 and q_2 are increased between $10^{-1.5} \rightarrow 10^{2.5}$ and $10^0 \rightarrow 10^{7.75}$, respectively. Table 4 shows the controller gains and the location of the poles for both the power control loop and the SOC control loop. The poles of the SOC control loop are selected such that its speed is several times slower than the ramp rate calculator.

The system and control parameters of the dc/ac converter are adopted based on the conventional voltage source converter (VSC) and vector control method [36–38] to regulate the dc voltage at 600 V. This ensures exchange of power between the dc and ac system in a desired way. The synchronous generator is simulated using the method proposed in Reference [39].

Table 4. BES system parameters.

BES System:
V_b: 1000 V,   RUR/RDR: 30%/minute $\approx$75 W/s
V_{bus}: 600 V, ΔP_{max}: 15 kW, ΔSOC_{max}: 30%
Q_{min}: 5040 As,   Choose Q: 5400 As

Converter: L_b:10 mH, R_b:10 mΩ

Controller:
Switching Frequency: 10 kHz

Power Control:
With q_1: 316.22, q_2: 0.0001
Poles: $-789.6 \pm j665.9$
Gains: k_{1b}: -17.78, k_{2b}: 0.0263

SOC Control:
With q_1: 316.22, q_2: 5.62 $\times$ 10^7
Poles $-0.0015 \pm j0.0011$
Gains: k_{1soc}: 17.7828, k_{2soc}: $-15{,}757$

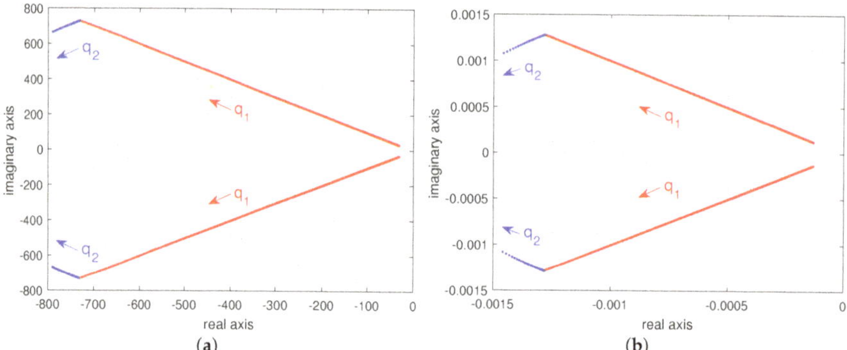

Figure 11. Poles of (**a**) the power, and (**b**) the state-of-charge (SOC) loops of BES.

4.1.2. Results and Discussion

The simulation scenario to study the impact of abrupt PV disturbances is defined as follows. The solar intensity is initially at 1000 W/m^2 and the system is operating at steady state. The PV system supplies 14.2 kW power and there is 1.4 kW power from the ac to the dc system. At $t=5.0$ s, the solar intensity drops from 1000 W/m^2 to 250 W/m^2 in a fraction of a second that drops the PV power to 3.42 kW. At $t=50$ s, the solar intensity goes back abruptly to 1000 W/m^2.

Figure 12 shows the output power response of the system. There is a sudden change in the power injected to the dc system (and then flowing into the ac system) when BES is not used. The BES smoothes down those changes and establishes an output power ramp at the desired rate. Figure 13 shows the transient response details of the PV, the BES, and the combined output powers. It confirms that the BES quickly compensates for the PV power practically without a delay.

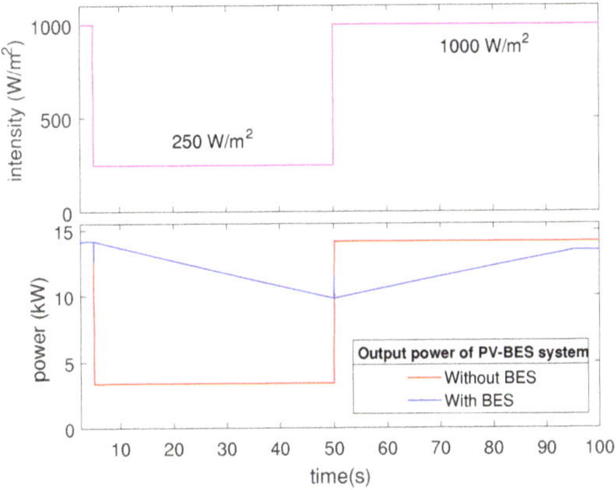

Figure 12. Top: solar irradiance, Bottom: powers of the PV and BES system response to the solar irradiance.

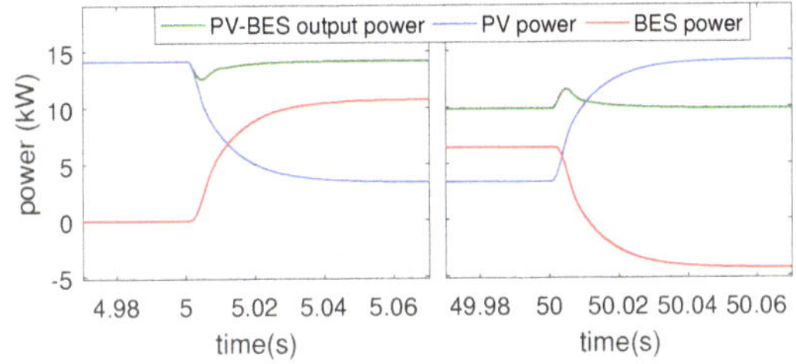

Figure 13. Power transients of the PV-BES system to PV disturbances.

Figure 14 shows the voltage and frequency transients at point of common coupling (PCC) of the dc and the ac system in response to the abrupt PV disturbances. The BES has reduced the peak and duration of both dc and ac voltage transients. The ac voltage (rms) experiences a peak fluctuation of close to 5.5 V (close to 5%) without the BES, and the BES limits the fluctuations within 0.1 V (practically zero). The BES has also improved the transients in the system frequency. Figure 15 shows the voltage transients at different buses in the dc system. The BES reduces the voltage transients across the entire dc system. Without the BES, this disturbances causes voltage fluctuations with peaks of over 10%. The BES reduces the peak of the transients to about 1%, and damps out the oscillations quickly. Over ten times improvement is achieved.

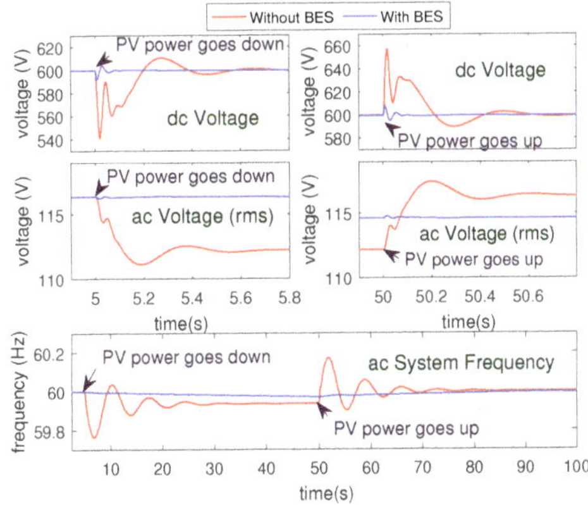

Figure 14. Voltage and frequency due to the abrupt PV disturbances.

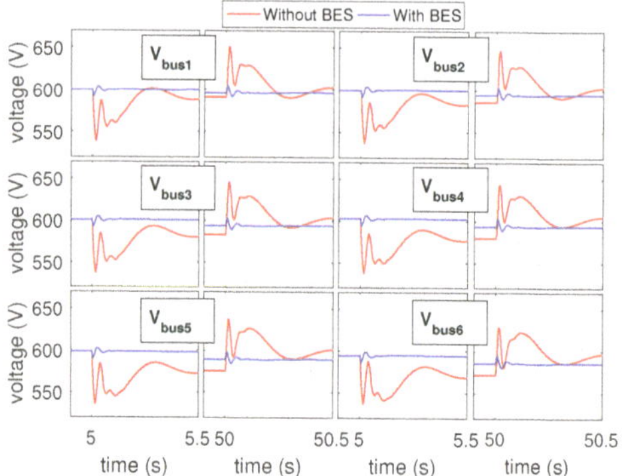

Figure 15. Dc system voltages due to the abrupt PV disturbances.

The ac current of the grid dc/ac converter is shown in Figure 16. When the PV power drops, the ac system supplies the deficit power to the dc system in the absence of the BES. With the proposed BES, the ac current has smooth transitions. Figure 17 shows the SOC of the BES in response to the abrupt PV disturbances. The SOC controller slowly regulates the SOC to the reference value after the disturbance is taken care without affecting the ramping behavior of the BES.

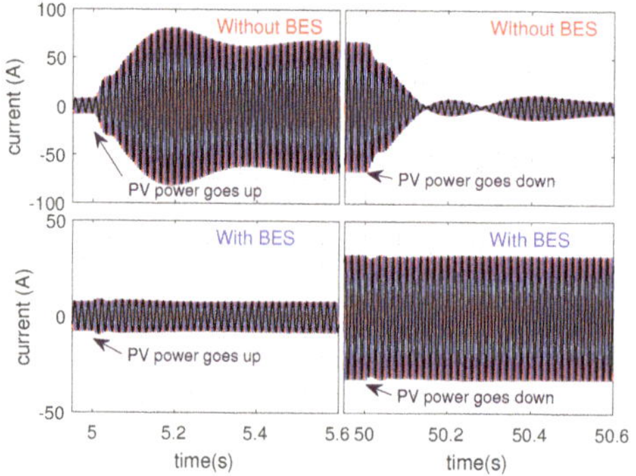

Figure 16. Inverter's ac current due to the abrupt PV disturbances.

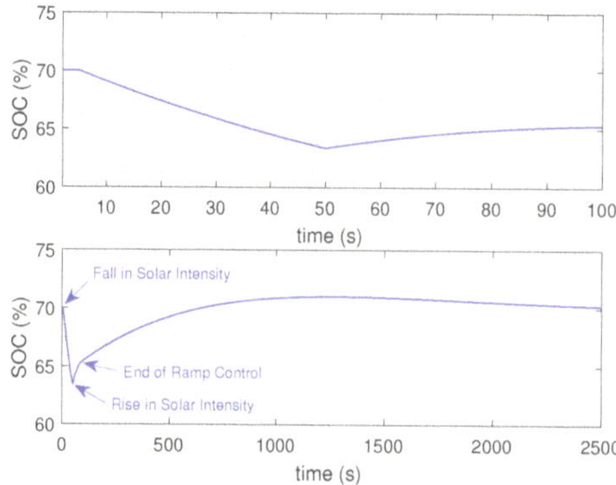

Figure 17. SOC of the BES during and after the abrupt PV disturbances.

4.2. Abrupt Disturbances: Case Study 2

Figure 18 shows the total power of the proposed PV-BES system (designed in previous section) in response to the real solar intensity measured at Lowry Range Solar Station on 18 March 2013 [31] for different power ramp settings (0.5%/minute, 1%/minute, 5%/minute). It demonstrates that the BES reduces the impact of the abrupt PV disturbances and smooths the power profile with the desired ramp rate.

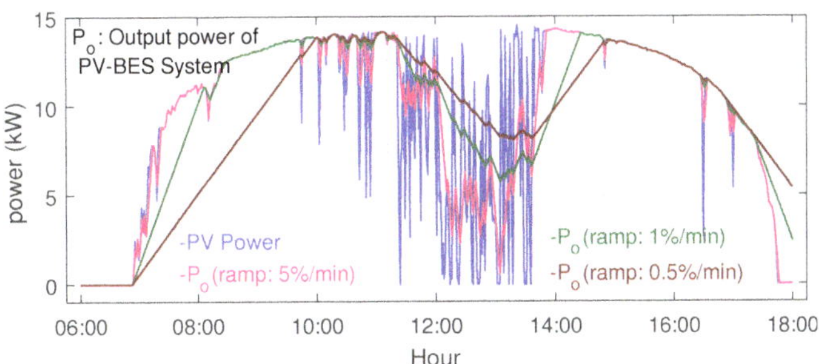

Figure 18. Output power of the proposed PV-BES system for different ramp settings in response to the real solar irradiance data.

4.3. Daily Power Profile of PV

The ramping capability of the proposed PV-BES system can be utilized to shape the daily load profile. When the distributed PV generators are equipped with the proposed BES system, and properly commanded, the aggregated load profile can become much smoother with some desired ramp rates. This also enables peak-load reduction. To demonstrate these properties, two scenarios are studied in this section as follows.

Scenario I: Figure 19 shows the power profile of California ISO on 31 March 2019 [33] with the proposed PV-BES system. The BES charges and stores energy in the morning as the solar intensity increases and avoids sharp decline of net load. This stored energy is released in the evening when the PV generation declines and the load demand increases. Figure 19 shows the profile when both RUR and RDR of the PV-BES system are set at 1200 MW/h. With a total of 5000 MW, 19411 MWh BES systems, the ramp rate is reduced from 5300 MW/h to 2767 MW/h (almost 50%) and the peak load is reduced from 22580 MW to 18747 MW.

Scenario II: In Figure 20, the RUR is set at 1400 MW/h while the RDR is set at 1000 MW/h till 20:00 and changed to 3000 MW/h afterwards. The required BES is 5571 MW, 15201 MWh and the ramp rate of the net load is reduced to 2660 MW/h with peak load of 21535 MW.

The results are summarized in Table 5. This study concludes that by properly adjusting the RUR and RDR, both the steep ramp-up rate and the peak of the net load profile may be mitigated. The RUR and RDR may be optimally computed and supplied by a secondary controller through a low-bandwidth communication.

Table 5. Impact of ramp rate settings on daily power profile.

S.N	Ramp Setting	BES Size		Maximum Ramp in Net Load	Peak Net Load
		Power	Energy		
1	No BES	-	-	5300	22,580
2	RUR: 1200 MW/h RDR: 1200 MW/h	5000 MW	19,411 MWh	2767 MW/h	18,747 MW
3	RUR: 1400 MW/h RDR: 1000 MW/h till 20:00 RDR: 3000 MW/h after 20:00	5571 MW	15,201 MWh	2660 MW/h	21,535 MW

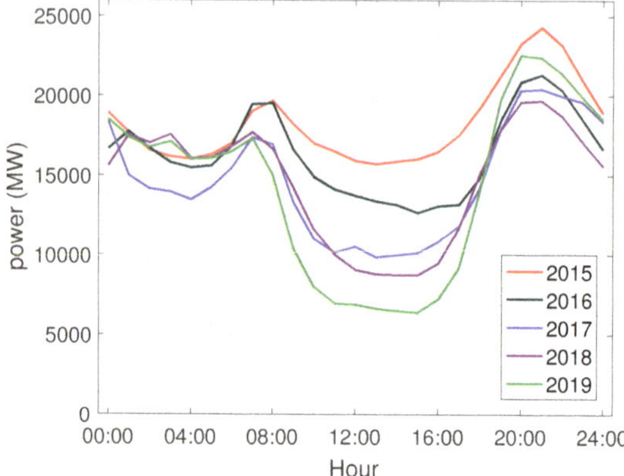

Figure 19. Shaping of Daily Load Profile: Scenario I.

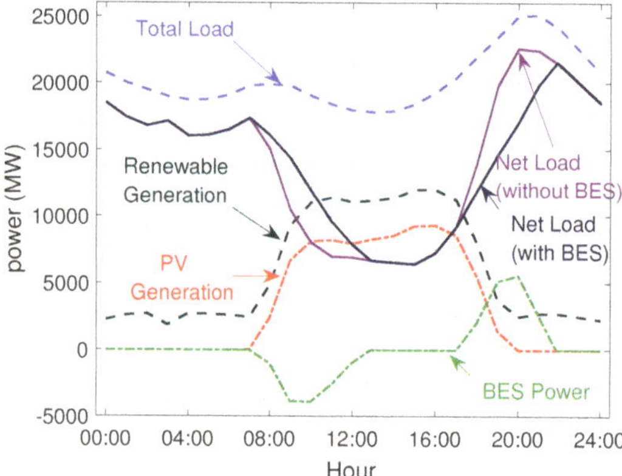

Figure 20. Shaping of Daily Load Profile: Scenario II.

4.4. Experimental Results

This section presents the results of a laboratory-scale implementation of the proposed system to evaluate its performance in a basic condition where all algorithms and controls are implemented in real time. The ability of the proposed approach to counter the PV disturbances (generated using an actual PV emulator hardware and other periphery circuits) is demonstrated. The results confirm feasibility of the proposed algorithms and controls. The details are presented throughout the section.

4.4.1. Experimental Setup

Figure 21 shows a low-power laboratory-scale experimental setup built to evaluate the performance of the proposed controller in response to the abrupt PV disturbances. An Agilent E4360A modular solar array simulator is used to emulate the PV and its disturbances. It is connected to a dc grid via a standard buck converter. A dc power supply in parallel with a local load models the dc grid. Another dc supply with a local load acts as a battery, and it is connected to the dc grid via a half-bridge converter for bi-directional power flow. The proposed control method is used and realized on the micro-controller to control the two converters.

Table 6 shows the system and control parameters. The control parameters for the PV system are designed to regulate the PV voltage to 20 V (maximum power point) using the method discussed in Section 4.1.1. For the BES system, the reference power calculation and the design of the power control parameters are done using the method in Section 3.3. We have ignored the SOC control and the ac system due to limitations in experimental setup.

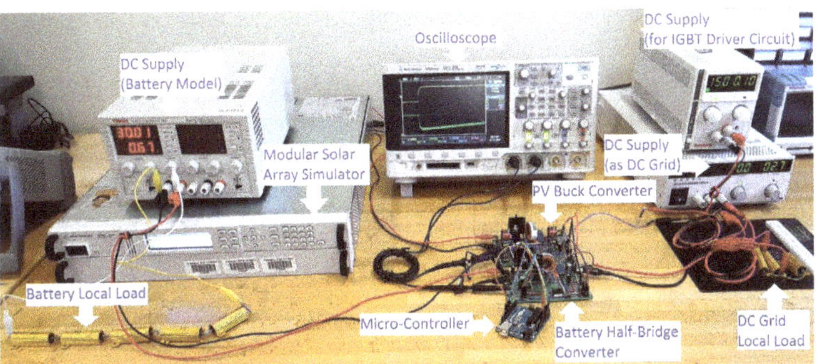

Figure 21. Experimental setup.

Table 6. Experimental setup parameters.

Modular Solar Array Simulator: **PV Curve I**: V_{mp}: 20.0 V, I_{mp}: 0.3 A, V_{oc}: 23.0 V, I_{sc}: 0.5 A   **PV Curve II**: V_{mp}: 20.0 V, I_{mp}: 2.0 A, V_{oc}: 23.0 V, I_{sc}: 2.5 A
PV converter: L_{pv}: 1 mH, R_{pv}: 0.2587 Ω, C_{pv}: 100μF
BES and converter: V_b: 30.0 V, Local Load: 30 Ω, L_b: 1 mH, R_b: 0.250 Ω
Dc bus: V_{bus}: 10.0 V, Local Load: 2 Ω
Controller: Switching Frequency: 32 kHz   PV converter control gains $k_{1\text{pv}}$: −56.23, $k_{2\text{pv}}$: 0.18, $k_{3\text{pv}}$: 10.00, $k_{4\text{pv}}$: 382.76, $k_{5\text{pv}}$: 1.02, $k_{6\text{pv}}$: −0.16   BES controller settings   RUR & RDR: 1.0 W/s, Control Gains: k_{1b}= −5.6234, k_{2b}= 0.0189 (Power control)

4.4.2. Results

Figure 22a shows the response of the system for the ramp rate setting of 1.0 W/s (for both RDR & RUR). Abrupt PV disturbances are manually applied to the PV emulator by switching from PV Curve I to PV Curve II and vice versa as specified in Table 6. When the PV conveter output increases, the battery quickly intervenes to absorb the power. Similarly, the battery supplies power when there is abrupt fall in PV converter power. With the grid voltage of 10 V, the ramp rate of 0.1 A/s in current is obtained. Figure 23 shows the results when the PV power experiences random abrupt disturbances. The study presented in this section confirms that the proposed control system can be successfully implemented in real-time on a typical hardware platform and the desired ramping rates set by the user are achieved by the combination of PV and proposed BES control system.

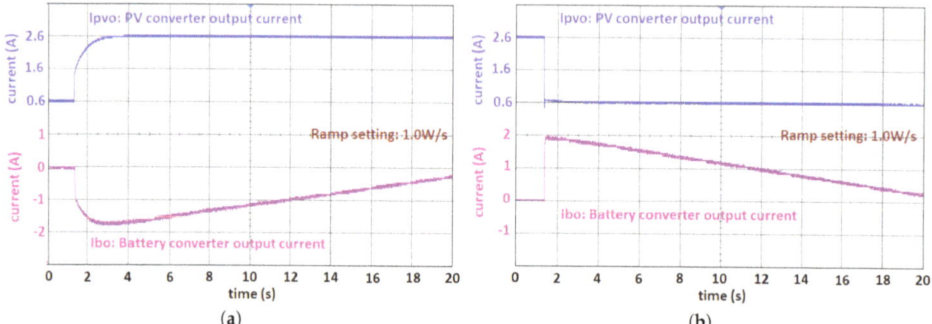

Figure 22. Experimental results: (**a**) ramp up response, and (**b**) ramp down response.

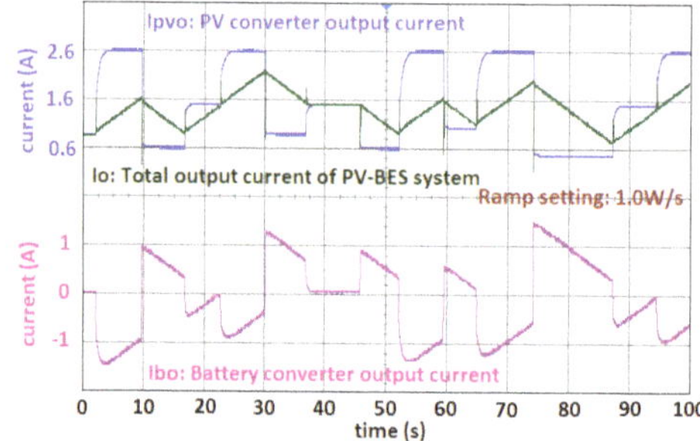

Figure 23. Experimental result: ramp up/down response for random PV power changes.

4.5. Comparisons

This section demonstrates comparison of the proposed method with conventional methods to execute smoothing of PV power. The comparison is done based on various qualitative parameters such as ramp calculation simplicity, plug-play level of proposed BES, coverage of converter model details, optimal control design, and inclusion of SOC control. The evaluation are done at the levels of poor, fair, average, and good. The comparisons are tabulated in Table 7.

Table 7. Qualitative comparison of proposed method with existing methods.

References	Smoothing Method	Ramp Calculation Simplicity	Plug-and-Play Level	Model Details	SOC Control Design	Optimal Control Design
[24]	Ramp	Average	Poor	Fair	No	No
[22]	Ramp	Fair	Poor	Fair	No	No
[25]	P-f droop	Average	Fair	Average	No	No
[26]	Ramp	Fair	Poor	Average	No	No
[27]	Ramp, Constant power	Fair	Fair	Average	No	No
[28]	Constant power	Average	Fair	Average	No	No
[29]	Constant power	Average	Poor	Average	No	No
[30]	Moving average, P-f droop	Fair	Poor	Poor	No	No
Proposed method	Ramp	Good	Good	Good	Yes	Yes

5. Conclusions

PV power experiences abrupt changes due to the fast moving clouds which cause voltage and frequency fluctuations in the weak grid and high PV penetrations. A battery energy storage (BES) is proposed in this paper to address fast power disturbances. Furthermore, the daily profile of the solar irradiance creates sharp ramp in daily load profile, which imposes challenges on the slow conventional generators. The ramping capability of the proposed PV-BES system is utilized to smooth and improve the daily load profile in terms of its steep ramp up and peak demand. Although the concept is developed for a BES system, it can be applied to other behind-the-meter distributed assets such as small gas turbine generators. The optimal computation and communication of the storage ramp rates are the future directions of research. To mention a limitation, the proposed approach of this paper requires the BES to be co-located with the PV and have access to its terminals in order to achieve the desired power ramping of the combined PV-BES.

Author Contributions: Conceptualization, R.S. and M.K.-G.; Modeling and simulation, R.S.; Formal analysis, R.S. and M.K.-G.; Investigation, R.S.; Writing–original draft preparation, R.S. and M.K.-G.; Writing–review and editing, R.S. and M.K.-G.; Supervision, M.K.-G.; Funding acquisition, M.K.-G. All authors have read and agreed to the published version of the manuscript.

Funding: This research work was partially supported by National Science Foundation (NSF 1808368).

Conflicts of Interest: The authors declare no conflict of interest.

Abbreviations

PV	Photovoltaic	BES	Battery Energy Storage
DER	Distributed Energy Resources	LQR	Linear quadratic regulator
MPPT	Maximum power point tracking	PCC	Point of common coupling
PI	Proportional Integrator	pu	per unit
RDR	Ramp Down Rate	RUR	Ramp Up Rate
SG	Synchronous generator	SOC	State of Charge
ISO	Independent System Operator		

Symbols and Variables

t	Time	k	Sampling instant
Δt	Change in time	S_{sg}	Synchronous generator power
v_{sg}	Voltage of synchronous generator	H	Inertia constant
R_{tr}	Resistance of transformer	X_{tr}	Reactance of transformer
VR	Transformer voltage ratio	V_{dc}	Dc voltage at the PCC
V_{ac}	L-L rms ac voltage at PCC	L_{inv}	Filter inductance of dc/ac converter
R_{inv}	Filter resistance of dc/ac converter	C_{dc}	Capacitor on dc side of dc/ac converter
V_{bus}	Voltage at PV-BES connection	$V_{bus\,i}$	Voltage of ith bus
RUR	Ramp up rate	RDR	Ramp down rate
R_{min}	Minimum ramp rate	P_{pv}	PV power
P_{pvo}	Output power of PV converter	ΔP	Change in PV power
P_r	Desired output power	ΔP_{max}	Maximum change in PV power
P_o	Total output power of PV-BES system	L_{pv}	Filter inductance of PV converter
C_{pv}	Capacitor across PV terminals	R_{pv}	Filter resistance of PV converter
$k_{i\,pv}$	PV control i^{th} gain	i_{pv}	PV current
i_{pvo}	Output current of PV converter	v_{pv}	PV voltage
v_{pv}^{ref}	PV reference voltage	V_{mp}	PV voltage at maximum PV power
I_{mp}	PV current at maximum PV power	V_{oc}	Open-circuit voltage of PV
I_{sc}	Short-circuit current of PV	P_b^*	BES reference power
P_{soc}^*	Reference power for SOC control of BES	P_b	Battery Power
P_{bo}	Output power of BES converter	i_b	Battery current
i_{bo}	Output current of BES converter	d_b	Duty cycle for BES converter
u_b	Control input of BES converter	v_b	Battery voltage
V_b^*	Nominal voltage of battery	SOC	State of charge of BES
SOC_n	Nominal SOC of BES	ΔSOC	Change in SOC of BES
ΔSOC_{max}	Maximum change in SOC	L_b	Filter inductance of BES converter
R_b	Filter resistance of BES converter	Q	Battery capacity
Q_{min}	Minimum battery capacity	$k_{i\,b}$	BES power control i^{th} gain
$k_{i\,soc}$	i^{th} gain of SOC control	J	Cost function
q_i	i^{th} weighting factor in cost function		

References

1. Al-Hallaj, S. More than enviro-friendly: Renewable energy is also good for the bottom line. *IEEE Power Energy Mag.* **2004**, *2*, 16–22. [CrossRef]
2. Hu, J.; Li, Z.; Zhu, J.; Guerrero, J.M. Voltage Stabilization: A Critical Step Toward High Photovoltaic Penetration. *IEEE Ind. Electron. Mag.* **2019**, *13*, 17–30. [CrossRef]
3. Administration, U.E.I. Annual Energy Outlook 2019: With Projections to 2050. 2019. Available online: https://www.eia.gov/outlooks/aeo/pdf/aeo2019.pdf (accessed on 8 April 2019).
4. Kroposki, B.; Johnson, B.; Zhang, Y.; Gevorgian, V.; Denholm, P.; Hodge, B.M.; Hannegan, B. Achieving a 100% renewable grid: Operating electric power systems with extremely high levels of variable renewable energy. *IEEE Power Energy Mag.* **2017**, *15*, 61–73. [CrossRef]
5. Silwal, S.; Karimi-Ghartemani, M. On Transient Responses of a Class of PV Inverters. *IEEE Trans. Sustain. Energy* **2019**, *10*, 311–314. [CrossRef]
6. Qunais, T.; Sharma, R.; Karimi-Ghartemani, M.; Silwal, S.; Khajehoddin, S.A. Application of Battery Storage System to Improve Transient Responses in a Distribution Grid. In Proceedings of the 2019 IEEE 28th International Symposium on Industrial Electronics (ISIE), Vancouver, BC, Canada, 12–14 June 2019; pp. 52–57.
7. Mather, B.A.; Shah, S.; Norris, B.L.; Dise, J.H.; Yu, L.; Paradis, D.; Katiraei, F.; Seguin, R.; Costyk, D.; Woyak, J.; et al. *NREL/SCE High Penetration PV Integration Project: FY13 Annual Report*; Technical Report; National Renewable Energy Lab. (NREL): Golden, CO, USA, 2014.
8. Katiraei, F.; Aguero, J.R. Solar PV integration challenges. *IEEE Power Energy Mag.* **2011**, *9*, 62–71. [CrossRef]
9. Bueno, P.G.; Hernández, J.C.; Ruiz-Rodriguez, F.J. Stability assessment for transmission systems with large utility-scale photovoltaic units. *IET Renew. Power Gener.* **2016**, *10*, 584–597. [CrossRef]
10. Sakamuri, J.N.; Das, K.; Altin, M.; Cutululis, N.A.; Hansen, A.D.; Tielens, P.; Van Hertem, D. Improved frequency control from wind power plants considering wind speed variation. In Proceedings of the 2016 Power Systems Computation Conference (PSCC), Genoa, Italy, 20–24 June 2016; pp. 1–7.
11. Gevorgian, V.; Booth, S. *Review of PREPA Technical Requirements for Interconnecting Wind and Solar Generation*; Technical Report; National Renewable Energy Lab. (NREL): Golden, CO, USA, 2013.
12. IEEE Std 1547-2018 Standard for Interconnection and Interoperability of Distributed Energy Resources with Associated Electric Power Systems Interfaces. 2018; pp. 1–138. Available online: https://ieeexplore.ieee.org/abstract/document/8332112 (accessed on 8 April 2019).
13. Denholm, P.; O'Connell, M.; Brinkman, G.; Jorgenson, J. *Overgeneration from Solar Energy in California. A Field Guide to the Duck Chart*; Technical Report; National Renewable Energy Lab. (NREL): Golden, CO, USA, 2015.
14. Ghosh, S.; Rahman, S.; Pipattanasomporn, M. Distribution voltage regulation through active power curtailment with PV inverters and solar generation forecasts. *IEEE Trans. Sustain. Energy* **2016**, *8*, 13–22. [CrossRef]
15. Xue, Y.; Guerrero, J.M. Smart Inverters for Utility and Industry Applications. In Proceedings of the PCIM Europe 2015; International Exhibition and Conference for Power Electronics, Intelligent Motion, Renewable Energy and Energy Management, Nuremberg, Germany, 19–21 May 2015; pp. 1–8.
16. Bragard, M.; Soltau, N.; Thomas, S.; De Doncker, R.W. The balance of renewable sources and user demands in grids: Power electronics for modular battery energy storage systems. *IEEE Trans. Power Electron.* **2010**, *25*, 3049–3056. [CrossRef]
17. Teleke, S.; Baran, M.E.; Bhattacharya, S.; Huang, A.Q. Rule-based control of battery energy storage for dispatching intermittent renewable sources. *IEEE Trans. Sustain. Energy* **2010**, *1*, 117–124. [CrossRef]
18. Ceja-Espinosa, C.; Espinosa-Juárez, E. Smoothing of photovoltaic power generation using batteries as energy storage. In Proceedings of the 2017 IEEE PES—Innovative Smart Grid Technologies Conference-Latin America (ISGT Latin America), Quito, Ecuador, 20–22 September 2017; pp. 1–6.
19. Lujano-Rojas, J.M.; Dufo-López, R.; Bernal-Agustín, J.L.; Catalão, J.P. Optimizing daily operation of battery energy storage systems under real-time pricing schemes. *IEEE Trans. Smart Grid* **2016**, *8*, 316–330. [CrossRef]
20. Hernández, J.; Sanchez-Sutil, F.; Muñoz-Rodríguez, F. Design criteria for the optimal sizing of a hybrid energy storage system in PV household-prosumers to maximize self-consumption and self-sufficiency. *Energy* **2019**, *186*, 115827. [CrossRef]

21. Gomez-Gonzalez, M.; Hernandez, J.; Vera, D.; Jurado, F. Optimal sizing and power schedule in PV household-prosumers for improving PV self-consumption and providing frequency containment reserve. *Energy* **2020**, *191*, 116554. [CrossRef]

22. Alam, M.; Muttaqi, K.; Sutanto, D. A novel approach for ramp-rate control of solar PV using energy storage to mitigate output fluctuations caused by cloud passing. *IEEE Trans. Energy Convers.* **2014**, *29*, 507–518.

23. Sangwongwanich, A.; Yang, Y.; Blaabjerg, F. A cost-effective power ramp-rate control strategy for single-phase two-stage grid-connected photovoltaic systems. In Proceedings of the Energy Conversion Congress and Exposition (ECCE), Milwaukee, WI, USA, 18–22 September 2016; pp. 1–7.

24. Li, X.; Hui, D.; Lai, X. Battery energy storage station (BESS)-based smoothing control of photovoltaic (PV) and wind power generation fluctuations. *IEEE Trans. Sustain. Energy* **2013**, *4*, 464–473. [CrossRef]

25. Hernández, J.C.; Bueno, P.G.; Sanchez-Sutil, F. Enhanced utility-scale photovoltaic units with frequency support functions and dynamic grid support for transmission systems. *IET Renew. Power Gener.* **2017**, *11*, 361–372. [CrossRef]

26. Kim, N.; Parkhideh, B. Control and Operating Range Analysis of AC-Stacked PV Inverter Architecture Integrated with Battery. *IEEE Trans. Power Electron.* **2018**, *33*, 10032–10037. [CrossRef]

27. Tran, V.T.; Islam, M.R.; Sutanto, D.; Muttaqi, K.M. Mitigation of Solar PV Intermittency Using Ramp-Rate Control of Energy Buffer Unit. *IEEE Trans. Energy Convers.* **2018**, *34*, 435–445. [CrossRef]

28. Saxena, N.; Hussain, I.; Singh, B.; Vyas, A.L. Implementation of a Grid-Integrated PV-Battery System for Residential and Electrical Vehicle Applications. *IEEE Trans. Ind. Electron.* **2018**, *65*, 6592–6601. [CrossRef]

29. Zhang, Q.; Sun, K. A Flexible Power Control for PV-Battery-Hybrid System Using Cascaded H-Bridge Converters. *IEEE J. Emerg. Sel. Top. Power Electron.* **2018**, *7*, 2184–2195. [CrossRef]

30. Rallabandi, V.; Akeyo, O.M.; Jewell, N.; Ionel, D.M. Incorporating Battery Energy Storage Systems Into Multi-MW Grid Connected PV Systems. *IEEE Trans. Ind. Appl.* **2019**, *55*, 638–647. [CrossRef]

31. Yoder, M.; Andreas, A. Lowry Range Solar Station: Arapahoe County, Colorado (Data). Available online: https://data.nrel.gov/submissions/30 (accessed on 8 April 2019).

32. Lew, D.; Miller, N. Reaching new solar heights: integrating high penetrations of PV into the power system. *IET Renew. Power Gener.* **2017**, *11*, 20–26, doi:10.1049/iet-rpg.2016.0264. [CrossRef]

33. California ISO. Renewables and Emission Reports. Available online: http://www.caiso.com/market/Pages/ReportsBulletins/RenewablesReporting.aspx (accessed on 8 April 2019).

34. Karimi-Ghartemani, M.; Khajehoddin, S.A.; Jain, P.; Bakhshai, A. Linear quadratic output tracking and disturbance rejection. *Int. J. Control* **2011**, *84*, 1442–1449. [CrossRef]

35. De Brito, M.A.G.; Galotto, L.; Sampaio, L.P.; e Melo, G.d.a.; Canesin, C.A. Evaluation of the main MPPT techniques for photovoltaic applications. *IEEE Trans. Ind. Electron.* **2013**, *60*, 1156–1167. [CrossRef]

36. Kazmierkowski, M.P.; Malesani, L. Current control techniques for three-phase voltage-source PWM converters: A survey. *IEEE Trans. Ind. Electron.* **1998**, *45*, 691–703. [CrossRef]

37. Blaabjerg, F.; Teodorescu, R.; Liserre, M.; Timbus, A.V. Overview of Control and Grid Synchronization for Distributed Power Generation Systems. *IEEE Trans. Ind. Electron.* **2006**, *53*, 1398–1409, doi:10.1109/TIE.2006.881997. [CrossRef]

38. Silwal, S.; Karimi-Ghartemani, M.; Sharma, R.; Karimi, H. Impact of Feed-forward and Decoupling Terms on Stability of Grid-Connected Inverters. In Proceedings of the 2019 IEEE 28th International Symposium on Industrial Electronics (ISIE), Vancouver, BC, Canada, 12–14 June 2019; pp. 2641–2646, doi:10.1109/ISIE.2019.8781454. [CrossRef]

39. Silwal, S.; Taghizadeh, S.; Karimi Ghartemani, M.; Hossain, J.; Davari, M. An Enhanced Control System for Single-Phase Inverters Interfaced with Weak and Distorted Grids. *IEEE Trans. Power Electron.* **2019**, *34*, 12538–12551, doi:10.1109/TPEL.2019.2909532. [CrossRef]

Article

Why PV Modules Should Preferably No Longer Be Oriented to the South in the Near Future

Riyad Mubarak *, Eduardo Weide Luiz and Gunther Seckmeyer

Institute for Meteorology and Climatology, Leibniz Universität Hannover, Herrenhäuser Straße 2, 30419 Hannover, Germany; luiz@muk.uni-hannover.de (E.W.L.); seckmeyer@muk.uni-hannover.de (G.S.)
* Correspondence: mubarak@muk.uni-hannover.de

Received: 7 October 2019; Accepted: 25 November 2019; Published: 28 November 2019

Abstract: PV modules tilted and oriented toward east and west directions gain gradually more importance as an alternative to the presently-preferred south (north in the Southern Hemisphere) orientation and it is shown to become economically superior even under the reimbursement of feed-in tariff (FIT). This is a consequence of the increasing spread between the decreasing costs of self-consumed solar power and the costs for power from the grid. One-minute values of irradiance were measured by silicon sensors at different orientations and tilt angles in Hannover (Germany) over three years. We show that south-oriented collectors give the highest electrical power during the day, whereas combinations of east and west orientations (E-W) result in the highest self-consumption rate (SC), and combinations of southeast and southwest (SE-SW) orientations result in the highest degree of autarky (AD), although they reduce the yearly PV Power by 5–6%. Moreover, the economic analysis of PV systems without FIT shows that the SE-SW and E-W combinations have the lowest electricity cost and they are more beneficial in terms of internal rate of return (IRR), compared to the S orientation at the same tilt. For PV systems with FIT, the S orientation presently provides the highest transfer of money from the supplier. However, as a consequence of the continuing decline of FIT, the economic advantage of S orientation is decreasing. E-W and SE-SW orientations are more beneficial for the owner as soon as FIT decreases to 7 Ct/kWh. East and west orientations of PV modules do not only have benefits for the individual owner but avoid high costs for storing energy—regardless who would own the storage facilities—and by avoiding high noon peaks of solar energy production during sunny periods, which would become an increasing problem for the grid if more solar power is installed. Furthermore, two types of commonly used PV software (PVSOL and PVsyst) were used to simulate the system performance. The comparison with measurements showed that both PV software underestimate SC and AD for all studied orientations, leading to the conclusion that improvements are necessary in modelling.

Keywords: incident solar radiation; PV output power modelling; tilt angle; orientation; rooftop solar

---

1. Introduction

Decarbonization of our energy supply is an important component to fulfill pledges of the Paris Agreement to keep the global warming below 1.5 °C, because 65% of the world's current CO_2 emissions are due to burning fossil fuels [1]. Renewable energy is one of the most cost-effective options to replace fossil fuels and to reduce electricity-related emissions. In recent years, many countries have begun a transition to more sustainable energy supply based on renewable energies. Solar energy represents the most abundant natural energy resource on the earth and has the potential to replace fossil fuels to satisfy this clean energy demand of our society in future [2]. This exceptional energy source is the most simple and economic renewable energy technology available that can be easily installed, especially on rooftops of houses. The costs for solar modules, measured in \$/Wp, have reduced by as much as 90% during the last decade and are expected to fall further in the future [3].

Consequently, the evolution of renewable energy over the past decade has surpassed most expectations. By the end of 2018, global total renewable generation capacity reached 2351 GW. PV solar electricity has developed rapidly in minor private systems, as well as in large-scale installations connected to national grids. Solar energy represented around 20.6% of renewable energy generation in 2018, with capacities of 486 GW [4].

The solar irradiance changes with geographical location, season, and time of the day according to sun position in the sky. In addition, it varies by the influence of clouds, aerosols, and ground reflection. The orientation and tilt angle of PV collectors are among the most important parameters that affect the performance of a PV system, as they determine the amount of solar radiation received by the PV collector [5]. The orientation and inclination of a PV installation has two effects on system output: On the one hand, there is a larger or smaller amount of total annual yield; on the other hand, there is an impact on the seasonal or daily timing of peak energy generation [6]. In general, PV systems are divided into fixed and tracking systems. Fixed systems are often small systems installed on the roof of a building, while tracking systems are often large PV systems installed to maximize the solar radiation that reaches them [7]. Module performance is also affected by local factors for individual locations e.g., cloudiness, temperature, shading, dust, precipitation, and bird droppings [8].

Based on Earth-sun geometry, many studies were carried out to find the optimum tilt angle and orientation of PV systems in certain areas worldwide, e.g., Italy [9], Turkey [10], Australia [11], the United States [12], India [13], China [14], and Ghana [15]. Most previous studies show that the optimal fixed tilt angle of PV collectors depends only on geographical latitude (φ), if local weather and climatic conditions are not considered. However, because of the diffuse solar radiation, the optimal tilt angles may differ from those in reality. Huld et al. [16] showed that climate characteristics have a huge influence on the optimal tilt angle in Europe. Lave and Kleissl [12] showed that the optimal tilt is reduced by up to 10 degrees when cloudiness is taken into consideration, particularly in the northern United States. European studies [17,18] concluded that the optimum tilt must be reduced by 10° to 20° between southern and northern Europe because of the same effect. Beringer et al. [19] showed that solar collectors oriented to the South at a tilt angle of 50°–70° in the winter months (October–March) and 0°–30° in the summer months (April–September) would result in the highest monthly yield for the location of Hannover, Germany.

Rooftop PV systems have gained importance in the last decade, especially from the drop in the cost of solar PV modules and the increase of end-consumer electricity tariff. According to recent studies, up to 25% of EU electricity consumption could be potentially produced in small rooftop PV systems installed in the existing EU building [20]. Other authors estimate that all electricity needs can be produced on rooftops [21]. There is increased interest in the self-consumption (SC), i.e., the part of PV power production that is consumed by the house owner. The savings from self-consumed PV-generated electricity are much higher than the profit from selling excess generation at spot prices. It may also have a positive effect on the distribution grid and make the production profiles of PV systems connected to the grid smoother.

The SC depends mainly on the system size: The more PV power installed, the more often the produced electricity exceeds consumption; i.e., it is non-linear with installed power [22]. SC can also be increased by energy storage and by load management; i.e., the influence of temporal resolution becomes less distinct with added a battery storage [23]. In practice, the SC rate can range from a few percent to a theoretical maximum of 100%, depending on the PV system size and load profile. Moreover, estimation of SC depends also on time resolution; i.e., it is overestimated when using hourly data of PV electricity production and household load profiles. Luthander et al. [24] found that for individual buildings, sub-hourly data are needed to capture the behavior of high peak power. Leicester et al. [25] found that SC is overestimated by 71.3% when using hourly data, compared with 54.8% when using one-minute data. Accordingly, high temporal resolution data are required to quantify SC accurately.

There are very limited studies that described simultaneous direct measurements of PV generation and consumption. However, one method to obtain more data with greater variety is to use PV data and separately-obtained load profile data, and estimate the SC fraction [25]. With the present reimbursement for feed-in tariffs that value just the yearly sum fed into the grid, suitability studies focused for rooftop have just concentrated on the yearly yield. Many studies and online web tools concerning the suitability of the orientation of rooftop implicitly take only the yearly sum into account [26]. Calculations for the diurnal variability are lacking.

In this study, we use one-minute data to compare the outputs of 12 solar collectors at various tilt and azimuth angles in order to propose an alternative concept for increasing SC via non-south-oriented PV systems and investigate its potential. The calculations are based on measurements from silicon sensors with different orientations and tilt angles in Hannover (Germany). The SC of all orientations is calculated by using a set of separately measured load profiles in order to evaluate the best and more-economic orientations for rooftop PV systems. The results are also compared with the simulated values of two widely used PV software packages, PVSOL [27] and PVsyst [28] to validate this software. Detailed information about the simulation parameters are listed in Tables.

2. Methodology

The input dataset used in this study is composed of one-minute output of 12 solar collectors (Figure 1) installed for three years (2016–2018) on the roof of the Institute for Meteorology and Climatology (IMUK) of the Leibniz Universität Hannover (Hannover, Germany; 52.23° N, 9.42° E and 50 m above sea level). Measurements have been made, using crystalline silicon PV devices with individual temperature sensors (Mencke and Tegtmeyer GmbH, Hameln, Germany). The PV devices have been calibrated by the manufacturer in November 2013 and they are cleaned regularly to prevent the accumulation of dirt and dust. In addition, all devices are compared after one year of measurements by placing them side by side horizontally. These comparisons were performed under different weather conditions and have showed an agreement within ±3%.

Figure 1. Set of solar PV devices based on silicon sensors mounted in several different tilt angles and orientations, operational at the IMUK (Institute for Meteorology and Climatology) [IMUK, 2017].

Two groups of identical devices are considered here: The first group consist of devices with 45° tilt, oriented to S, E, W, SE, and SW; the second group consists of vertical devices, oriented to S, E, W, SE, SW, and N. The tilt angle for the first group (45°) is chosen to represent the large number of roof pitches, where most residential houses in Germany were built with a tilted roof angle between 40° and 45° [29]. According to the initial design of the measurement system, the S measurements are conducted at tilt angles of 40° and 50°, therefore we take the average of both sensors (40° and 50°) to represent the PV outputs at 45° tilt; the uncertainty resulting from this procedure and other orientation uncertainties are less than 1% according to PVGIS calculations. Table 1 shows an overview about the inclination uncertainty, according to a Photovoltaic Geographical Information System (PVGIS) calculation [30] for Hannover.

Table 1. Annual PV energy produced in Hannover with respect to the optimal inclination [%] according to PVGIS.

East				←				Azimuth				→					West
tilt	90	80	70	60	50	40	30	20	10	0	10	20	30	40	50	60	70	80	90
0	86	86	86	86	86	86	86	86	86	86	86	86	86	86	86	86	86	86	86
10	86	87	88	89	90	91	92	92	93	93	93	92	92	91	90	89	88	87	86
20	84	86	88	90	92	94	95	96	97	97	97	96	96	94	93	91	89	87	84
30	81	84	87	90	93	95	97	98	99	99	99	98	97	96	93	91	88	85	81
40	78	82	85	89	92	94	96	98	99	100	99	98	97	95	92	89	86	82	78
50	74	78	82	86	89	92	94	96	97	97	97	96	95	92	90	87	83	79	74
60	70	74	78	82	85	88	90	92	93	93	93	92	91	89	86	82	79	74	70
70	65	69	73	76	80	82	84	86	87	87	87	86	85	83	80	77	73	69	65
80	59	63	67	70	73	75	77	78	79	80	79	79	78	76	73	70	67	64	60
90	53	57	60	63	65	67	69	69	70	70	70	70	69	68	66	63	60	57	54

3. PV System Output Calculation

In general, there are several ways to calculate the power output of PV systems. We used in this study a simple method for calculating it [31]:

$$P_{m,i} = P_{rel} \times \frac{I_{m,i}}{I_{UTC}} \times \left(1 + \gamma\left(T_{sen,i} - 25\,^{\circ}C\right)\right) \times PLF \tag{1}$$

where P_m is power output of the PV system, P_{rel} is the rated PV system power (the output power of PV device under standard test conditions), I_m is the measured solar irradiance, $I_{UTC} = 1000$ W/m^2, T_{sen} is the module temperature (in °C), γ is power temperature coefficient, and PLF is the power loss factor.

The equation contains the temperature coefficient to take into account the drop of sensor signal because of the temperature and to correct the testing conditions. The losses because of inverter and the degradation mechanisms of the PV sensors (0.5%/a) are included in Equation (1) as a PLF, which is time dependent because of the degradation of sensors.

3.1. Load Profile

The power generation profiles were calculated by using the Equation (1). A synthesized dataset of actual measured load profiles provided by HTW Berlin [32] is used to simulate a household's consumption pattern of electricity. The data set consists of 74 load profiles of German single-family houses with a temporal resolution of 1 min for every day of the year. The load profile used for the calculations is the average of six selected profiles which have an annual consumption between 3900 kWh and 4055 kWh. The average profile has an annual electricity consumption of 4006 kWh (Figure 2). It can be assumed that the selected profiles represent a four-person household.

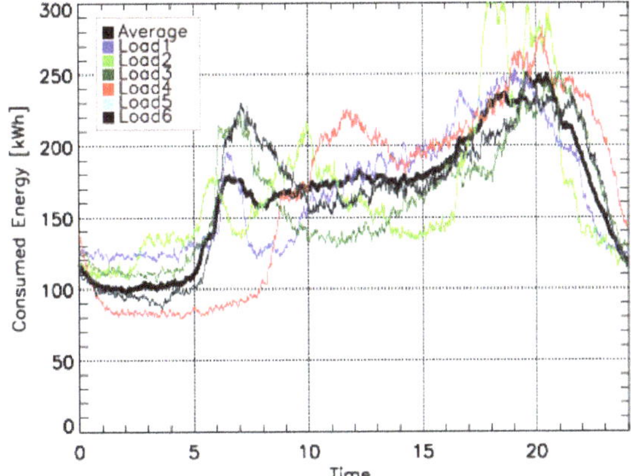

Figure 2. Six private household profiles which have an annual consumption between 3900 kWh and 4055 kWh [32]. The average profile (black curve) has an annual electricity consumption of 4006 kWh.

3.2. Economic Parameters

Feed-in tariffs are the most common policy instrument worldwide to support renewable energy. Many PV installations sell their power at local grid, and the majority of feed-in tariff contracts are at a fixed price per kWh for 10–20 years [33]. This results in an optimal orientation that is the same for both maximum economic yield and maximum energy production. The German FIT for solar photovoltaic uses varying rates depending on the size of the project. Countries in which the FIT was eliminated usually replace it by net metering schemes. The net metering is also used in many different countries under different rules, but consists of a system in which the excess electricity injected into the grid can be used at a later time to compensate the consumption when PV generation is not sufficient. The compensation usually covers a specific period (usually 1–3 years) depending on the country's regulations, and any excess energy after this period is not remunerated. So, the main idea is to configure the system settings in a way its annual production does not exceed the annual consumption, minimizing the deviation between them and increasing SC. Examples of countries using net metering schemes are: the United States (with particular conditions depending on the state), Denmark, Greece, Australia, Brazil, Mexico, and Chile [34–36].

The FIT used in the financial model for the calculation is 10.64 Ct/kWh (from July, 2019) and the price is constant for 20 years. The electricity price (30.22 Ct/kWh) considered in the calculations in this study represents the average price level for private households in Germany in 2019, including taxes and levies [37]. The increase of electricity price is expected to slow down to 2% p.a. as an average value during the next 20 years. The levelized cost of PV energy (LC) in northern Germany ranges between 9.89 Ct/kWh and 11.54 Ct/kWh, depending on the annual solar irradiance [38]; a value of 10 Ct/kWh is used in this study.

In the design of PV systems, the self-consumption rate (SC) and the degree of autarky (AD) are two important quantities used to assess the congruence of the PV generation and electricity demand profiles. The self-consumption rate is defined by the ratio of PV directly used (P_{DU}) to the total amount of PV power generated (P_m), according to Equation (2).

$$SC = \frac{P_{DU}}{P_m} \tag{2}$$

The degree of autarky is defined as a ratio of PV directly used to the total consumption by the household [39], according to Equation (3).

$$AD = \frac{P_{DU}}{L} \tag{3}$$

where L is the energy consumed by the loads.

The electricity price P_E used to evaluate the economic impact of PV system at specific orientation has been calculated according to Equation (4).

$$P_E = (P_G - P_{Fi}) + LC \tag{4}$$

where P_G is the grid electricity price, P_{Fi} is the FIT, and LC is the levelized cost of PV energy.

Figure 3 shows a workflow diagram used in this study to calculate the SC with the feed-in components. The calculations are always dependent on the consumption of electricity, with the primary objective to fulfil the demand from the PV produced energy, before purchasing from the public grid. If the produced electricity exceeds the consumption of the house, the excess is supplied to the public grid. Moreover, the internal rate of return (IRR) for all available orientations has been calculated over the life cycle of the PV system (20 years) in order to enlighten prospective owners/investors of rooftop PV systems. The IRR, defined as a discount rate that makes the net present value from all cash flows from a project equal to zero, is used to evaluate the attractiveness of a project or investment, and it is probably one of the most meaningful metric for investors [40]. The degradation mechanisms of the PV collectors (0.5%/a) and an annual increase of electricity price (2%/a) were taken into account in the IRR calculations.

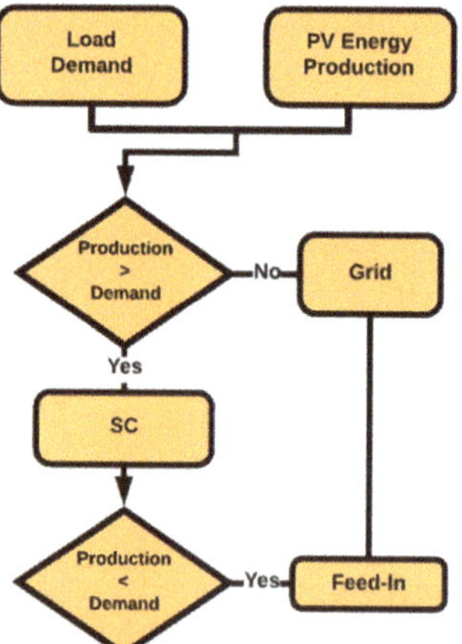

Figure 3. Schematic view of the calculation of system components. The calculations are always dependent on the load demand, with the primary objective to fulfil it from the PV produced energy, before purchasing electricity from the public grid.

3.3. PV Software

PV estimation models are generally used to estimate the expected energy output of a PV system. These models need specific input parameters such as meteorological conditions of the location, system design details, and definitions of the main components used. A variety of software for the simulation of PV systems is available in the market, including PVsyst, PVSOL, and others. PVsyst, developed at the University of Geneva, is one of the most common modeling software tools used in the PV industry to simulate the performance of grid-connected or stand-alone PV systems and calculate their energy yield.

PVsyst allows the definition of meteorological databases from many different sources and formats, as well as on-site measured data [41]. On the other hand, PVSOL is a German software developed by Valentine Software [27] for dynamic simulation with 3D visualization and detailed shading analysis of photovoltaic systems. PVSOL gives customers the best return on their investment by visualizing systems, and it can perform economic and performance analysis with comprehensive reports.

4. Results and Discussion

4.1. Production and Consumption under Different Weather Conditions

Figure 4 shows the PV production profiles for the S and E-W orientations during three days of May 2018 with different weather conditions (clear sky, partly cloudy, and fully cloudy) and correspondent customer load profiles for the same days of the year.

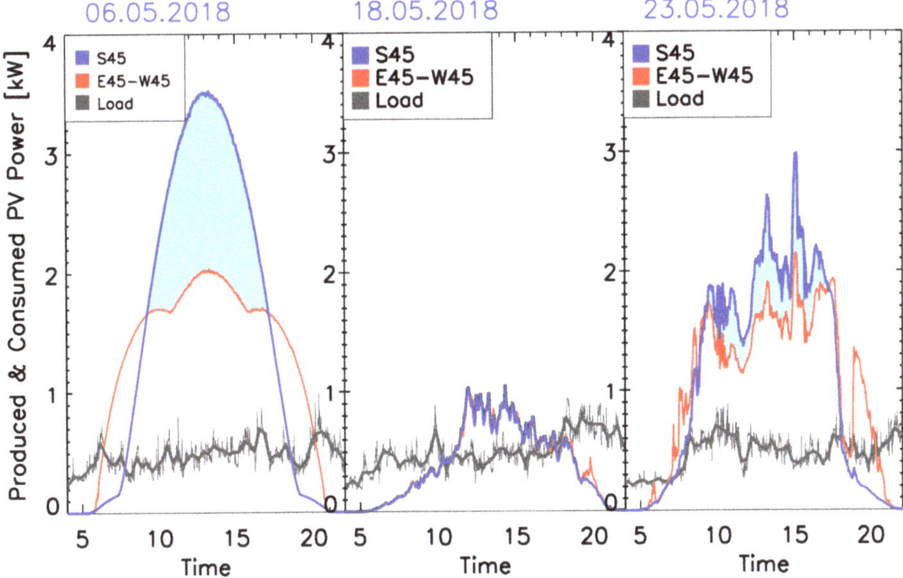

Figure 4. PV Production profiles for three days of May 2018 and the correspondent customer load profiles for the same days of the year. The thick (gray) curve represents the 10-minte average consumption.

The load profiles show different peaks over the day according to consumption patterns, while, the PV power production changes according to the movement of the sun and the weather conditions. The influence of orientation is shown clearly in clear sky days, when the energy production depends mainly on the sun's position. The E-W orientation covers more the edges of the day and reduces noon peak. On the other hand, the orientation is irrelevant under cloudy conditions, when the solar irradiance dominated by diffuse component. In general, SC rate is higher under cloudy conditions.

4.2. Annual Insolation

The annual total solar energy as function of surface azimuth and tilt angles is depicted in Figure 5. The left side histogram shows that the maximum annual total energy is for a south-facing surface with a tilt angle between 30° and 40°, closer to 40°. The annual total energy is less than the maximum by approximately 0.2% for surface orientation of 30° S, and it decreases gradually with higher or lower tilt angles. The annual produced energy for the surfaces oriented at the same tilt angle (45°) toward E and W are 77.0% and 75.9% of the optimal orientation respectively. For orientations of 45° SE or SW, the annual total energy produced are 94.8% and 93.3% of maximum produced energy. The inequality in total energy for E and W and in SE and SW may denote asymmetric distributions of solar irradiance before and after midday. On the right histogram of Figure 4 we can see that, for a vertical surface with orientation of 90° south, the produced energy is 66.2% of the 40° tilted surface, whereas it is about 50% for E and W surfaces. The annual total energy of the northern vertical surface is reduced by about 74%.

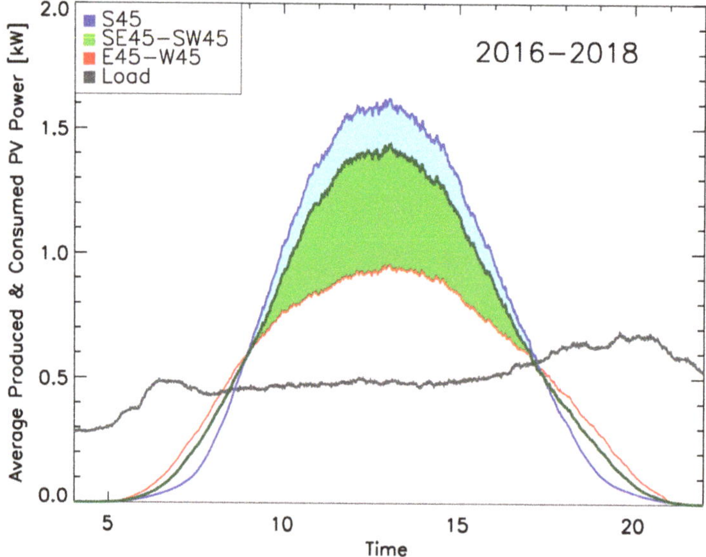

Figure 5. Produced and consumed PV power averaged over all days of the years 2016–2018 for the S, E-W, and SE-SW orientations at 45° tilt. The area below the gray curve represents the average load profile. The E-W and SE-SW facing installation produce more electricity in the mornings and evenings with a lower midday peak, so they match the load profile more closely.

In general, the amount of energy produced by a PV collector is proportional to solar radiation received by a surface in a specific orientation. Table 2 shows the annual produced energy and its percentage from the maximum value (at 45° S) for different orientations and tilt angles. The table also shows the SC rate and AD for each orientation. For 45° tilt surfaces, the lowest SC rate (37.9%) is for the S facing solar installation, while the highest SC rate (51.4%) is for the E-W combination. The high SC rate is because the power output of E-W installation matches the load profile more closely, producing more electric energy at the beginning and at the end of the day, with a lower midday peak (Figure 6). The AD has its maximum at SE-SW combination (40.7%) and its minimum at the E orientation (35.4%). Moreover, the economic efficiency of all studied orientations for the cases with and without FIT is also listed in Table 2. Overall, the E-W and SE-SW combinations have the lowest electricity cost (29.2 Ct/kWh and 29.1 Ct/kWh respectively), while the E orientation has the highest one (30.7 Ct/kWh), both cases for the system without FIT. For PV systems with FIT, the S-facing systems have the lowest electricity price (22.0 Ct/kWh) because of the high PV generation and accordingly the high feed-in amount, while the E-facing systems have the highest price (25.6 Ct/kWh).

Table 2. Results of measurement for a PV size of 4.8 kWp at different tilt angles in Hannover, Germany.

Orientation		S45	E45	W45	E45 + W45	SE45	SW45	SE45 + SW45	S90	E90	W90	SE90	SW90	N90
Percentage of SS45 (%)		100	76.2	75.0	75.6	94.6	93.1	93.8	66.8	49.8	50.3	64.6	64.4	25.8
Annual PV Generation (kWh/a)		4145	3157	3111	3134	3921	3859	3890	2769	2064	2046	2678	2655	1069
SC rate (%)		37.9	44.9	47.9	51.4	38.6	40.7	41.9	50.9	57.4	66.2	50.0	55.4	95.1
Autarky (%)		39.2	35.4	37.2	40.3	37.8	39.2	40.7	35.2	29.6	34.5	33.4	36.7	25.4
Cost Ct/kWh	No FIT	29.5	30.7	30.1	29.2	30.5	29.5	29.1	30.9	32.7	31.1	31.4	30.4	34.0
	With FIT	22.0	25.6	25.3	24.7	23.0	22.8	22.5	27.0	30.2	29.1	27.6	27.1	33.9
IRR % Over 20 ys	No FIT	1.59	0.20	0.84	1.90	1.10	1.57	2.09	0.13	−2.15	−0.19	−0.53	0.70	−4.23
	With FIT	7.05	4.13	4.42	5.14	6.29	6.45	6.82	3.23	0.12	1.45	2.61	3.33	−4.10

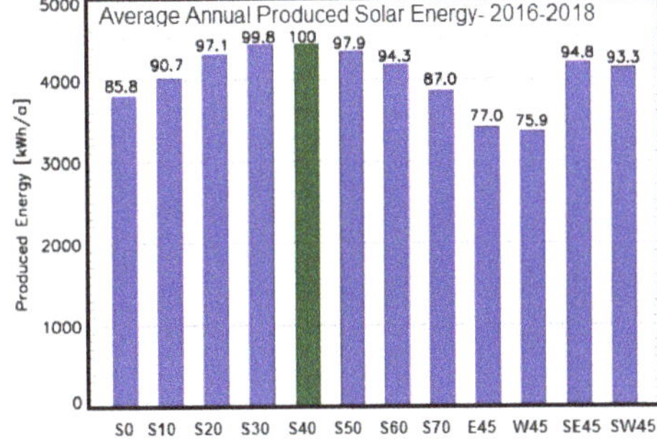

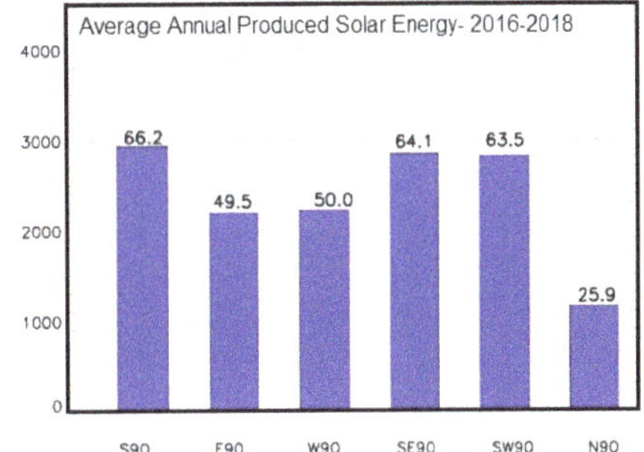

Figure 6. Average annual total solar energy (2016–2018) measured at IMUK and normalized values (in %) with respect to the annual total maximum energy at 40° S. The solar energy decreases for higher or lower tilt angles and for other azimuth angles.

The IRR analysis of PV systems without FIT shows that the SE-SW and E-W orientations tilted at 45° is more beneficial with an IRR value of 2.09% and 1.90%, respectively, when compared to the S orientation at the same tilt with 1.59%. For PV systems with FIT, the IRR for the S orientation is higher with a value of 7.05%, compared to the SE-SW and E-W orientations with 6.82% and 5.14%, respectively.

As expected, for the vertical surfaces, the S orientations gives the highest output (66.8% of the maximum), while the lowest energy is produced by N-facing surface (25.8 of the maximum), because of the Earth-sun geometry in the northern hemisphere. In terms of the SC rate, the N surfaces have the highest rate, due to the low energy production in this direction, while the lowest rate (50.0%) is for the SE surface. The AD has its maximum at SW orientation (36.7%) and it is minimum at N orientation (25.4%). Accordingly, the SW orientation has the lowest electricity cost (30.4 Ct/kWh) for the system without FIT, while the S and SW orientations have the lowest electricity cost (27.1 Ct/kWh) for PV systems with FIT. The difference between prices is found to be small and is within ±3%. However, we found that a changing the irradiance of 3% cause only a small change of the price and therefore conclude that the assessed measurement uncertainties do not significantly affect the prices.

Moreover, we examined whether the PV self-consumption will be influenced similarly in all investigated orientations, by changing the system size. For that purpose, we varied the module area by +/−50% in 5% steps (Figure 7). As expected, the SC rate increased by a reduction of the module area (in our specific case the default area was 24.3 m^2) for the orientations (S45, E45-W45, SE45-SW45). This increase only slightly depends on the orientation: The E40-W40 increased by 29% while the south orientation increased by 26% with a reduction of the module area of 50%. While it is obvious that the SC rate becomes smaller for larger module areas, an increase in module size will affect all orientations, but the S orientation will be affected slightly less than the other orientations.

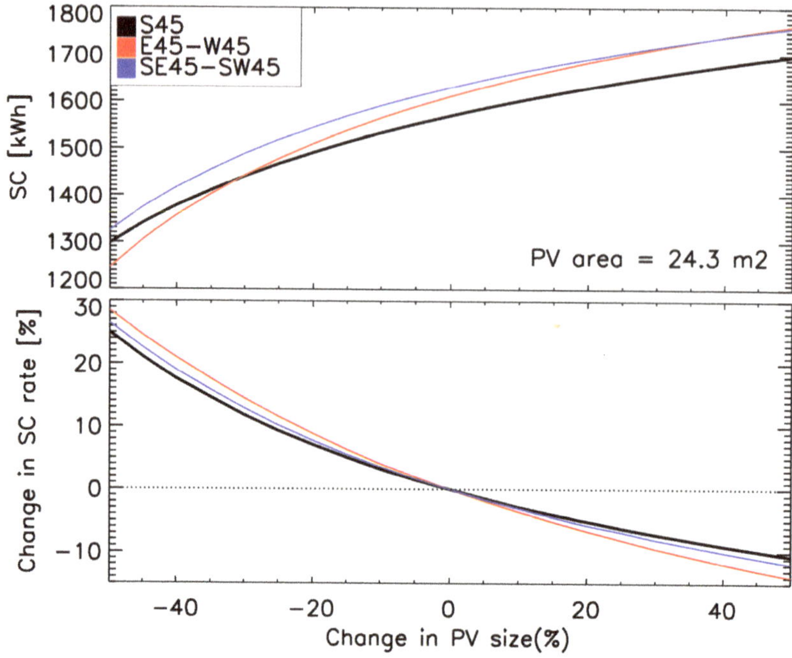

Figure 7. The change in self-consumption (SC) rate with varying the module area by +/−50% in 5% steps. The change in SC rate depends on the orientations.

4.3. Effects of the Changing Feed-In Tariffs

Feed-in tariffs of renewable energy in Germany are decreasing as each year passes and PV FIT drops faster than any other renewable power source. In the last 15 years, the FIT recorded a decrease of approximately 80% for small rooftop PV installations and 90% for medium-size PV systems [42]. Figure 8 shows the decrease in German FIT from 2000 to 2020.

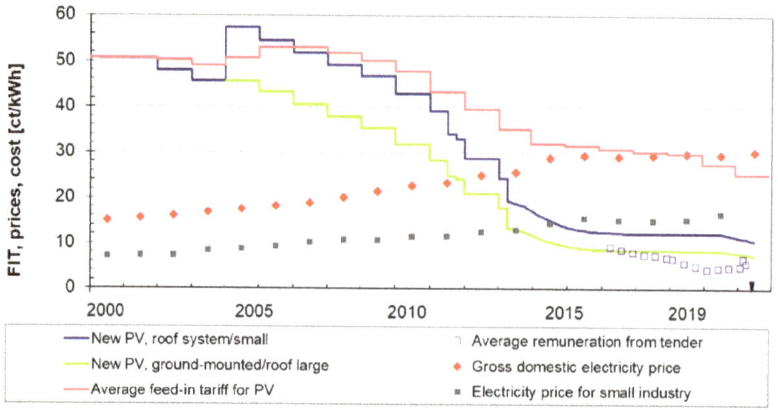

Figure 8. The changes of PV FIT and electricity price in Germany (2000 to 2020). The FIT dropped approximately 80% for small rooftop PV installations and 90% for medium-size PV systems [42].

According to Obane and Okajima [43], the FIT scheme for small PV systems is fast approaching its closure or expiration in many countries. In Germany, the EEG law stipulates that further FIT systems will not be allowed, when the total PV installations reach 52 GW. At the end of April 2018, the country had 43.8 GW PV installed. With the current tenders of PV, this cap is expected to be reached in 2020 [44]. However, the German government presently reconsidering this plan and is considering to allow FIT in future when the 52 GW is exceeded. With decreasing FIT self-consumption is gaining higher importance, especially with increasing cost of delivering PV electricity and rapid decline in the cost of solar PV modules. In addition, after 2020, the FIT will gradually expire for the oldest PV plants [42] and the produced PV power will be mainly used for SC. Thus, E-W and SE-SW orientations will provide the highest SC rate and will be more beneficial for the householders. Our calculation shows that the higher benefit of south orientation is no longer existent if FIT decreases to 7.0 Ct/kWh or lower, where at least the SE-SW orientation will have a higher IRR that the S orientation.

The economic calculations above were done for the conditions of the present FIT in Germany. The major conclusions, however, can used for many countries around the world, which apply FIT or similar PV cost structures. The results are especially relevant for the countries, which offer a very low FIT (e.g., New Zealand and Portugal) or for which eliminated the FIT scheme (e.g., UK, Spain, Czech Republic, Italy).

4.4. Comparison with PV Software

For simulation of the IMUK measurement system, a fixed PV system configuration, consisting of a 4.8 kWp is considered in the calculations, corresponding to the installation of 24 modules. Moreover, the same load profile that is used for the calculation of SC and AD at IMUK is also used in both models. Table 3 shows the important model parameters used in the simulation.

Table 3. Model parameters used in comparison.

Parameter	Model (PVSol, PVSyst)
Modules	4.8 kWp, mono, 24 modules
Inverter	ABB, 4.6 kW
Climate data	Meteonorm 7.2
Transposition model	Perez-Ineichen model
Diffuse radiation model	Perez model

Both simulation programs have been run for each orientation separately. Table 4 shows the simulated annually produced energies for all studied orientations and tilt angles. The programs overestimate the south-tilted irradiance and most of the studied orientations. This may result from the use of an anisotropic model (Perez-Ineichen model) to calculate the tilted irradiance, where we found in a previous study [45] that anisotropic models overestimate the south-tilted irradiance and most vertical-tilted irradiances.

The table also shows the SC and AD fractions for each orientation. For the 45°-tilt surfaces, the lowest SC (PVsyst = 32.9% and PVSOL = 32.3%) are for the S orientation, while the highest SC (PVsyst = 43.1% and PVSOL = 44.5%) is for the E-W combination, which agrees with the measured results. According to PVsyst, the AD has its maximum at E-W combination (37.4%) and at S orientation (34.7%) according to PVSOL calculations, while it is minimum at the W orientation for both models (PVsyst = 34.0%, PVSOL = 31.9).

For the vertical surfaces, the results of both programs show also that the S orientations gives the highest output, while the lowest energy is produced by a N-facing surface. In terms of the SC rate, the N surface has the highest fractions (PVsyst = 85.7% and PVSOL = 89.0%), while the lowest (40.7%) are for the S surfaces. The AD has its maximum for S surfaces (PVsyst = 31.8% and PVSOL = 30.6%) and it is minimum at the N orientation (PVsyst = 24.6% and PVSOL = 24.7). Table 4 also shows that both PV programs overestimate the percentage of energy production at 45° in most orientations versus the southern maximum value.

Table 4. Results of PV software PVSPL and PVSyst for a PV size of 4.8 kWp at different orientation and tilt angles.

Orientation	S45	E45	W45	E45 + W45	SE45	SW45	SE45 + SW45	S90	E90	W90	SE90	SW90	NN90
PVsyst Annual PV (kWh)	4457	3596	3531	3564	4250	4174	4212	3161	2439	2362	3053	2938	1148
Percentage of the max (%)	100	79.5	79.3	78.7	94.8	94.1	93.9	71.0	55.1	54.6	68.4	67.2	26.0
SC (%)	32.9	40.7	38.9	43.1	34.9	33.9	36.1	40.7	49.5	46.6	42.3	40.9	85.7
AD (%)	36.2	35.6	34.0	37.4	36.4	35.1	37.3	31.8	30.1	28.0	31.9	30.3	24.6
PVSOL Annual PV (kWh)	4330	3425	3046	3148	4160	3857	3920	3012	2289	1975	2945	2629	1115
Percentage of the max (%)	100	77.8	68.7	71.3	95.9	88.3	90.1	69.9	53.1	45.9	68.4	61.1	25.8
SC (%)	32.3	39.9	43.1	44.5	33.4	35.6	35.6	40.7	50.4	57.0	41.5	45.6	89.0
AD (%)	34.7	33.4	31.9	34.1	34.5	33.8	34.5	30.6	28.8	28.1	30.5	29.9	24.7

In order to have comparable results of simulation with the measured results, the generated PV energy of the IMUK system have been controlled by changing the PV area to produce the same annual output as the inverter output of simulation software. Figure 9 shows the results of the comparison: Both PV programs underestimate SC and AD for all studied orientations; SC rate was underestimated by 0.4% to 14%, while AD values were underestimated by 1.3% to 8.1%. These results lead to the conclusion that improvements are necessary in the modelling of SC and AD.

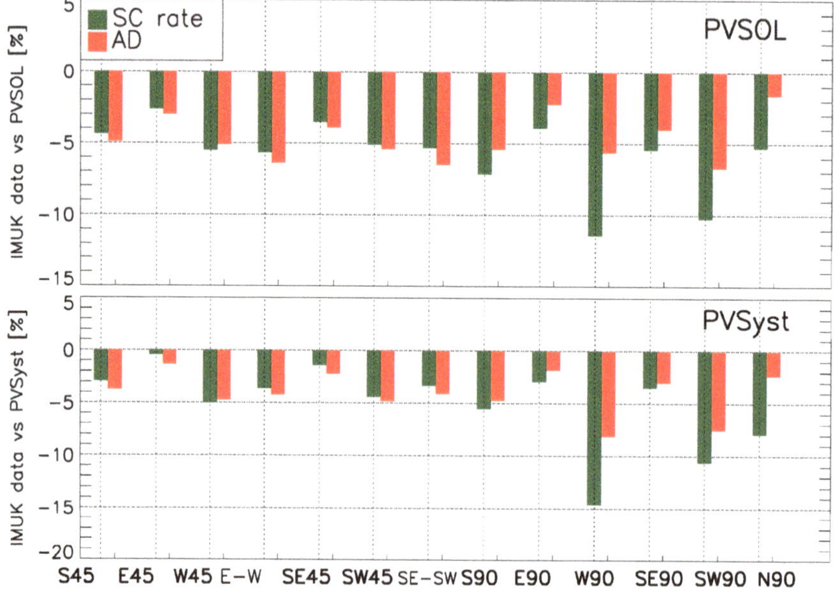

Figure 9. Comparison between IMUK results and simulated values. The used PV simulation software underestimate self-consumption and degree of autarky at all studied orientations and tilt angles.

5. Conclusions

Using one-minute measured data of PV energy, the outputs of 12 solar collectors at various tilt and azimuth angles in Hannover (Germany) were analyzed. For validation, the results were also compared with the simulated values of two widely used PV software: PVSOL and PVsyst.

The measurements show that a south-oriented generator at about 40° gives the highest electricity profile. For non-vertical devices, the combinations of E and W orientations result in the highest SC rate and combinations of SE and SW orientations result in the highest AD. E-W and SE-SW combinations have the lowest electricity coast for PV systems without FIT, while the E orientation has the highest one. For PV systems with FIT, S orientation provides the highest transfer of money from the supplier. The economic analysis using IRR of PV systems without FIT shows that the SE-SW and E-W orientations tilted at 45° is slightly more beneficial, while S orientation has higher IRR for PV systems with FIT.

However, in light of the continuing decline of FIT, the advantage of S orientation is decreasing and our results show that E-W and SE-SW orientations will be more beneficial if FIT is to 7 Ct/kWh or lower. East and west orientations of PV modules and not south orientations should be supported because they would also reduce the economic costs for storing renewable energy—regardless who would own the storage facilities—and avoid high noon peaks of solar energy production, which would become a problem for the grid for higher solar power penetrations levels.

Furthermore, the results show that the vertical tilted surfaces represent a high potential for PV energy production and facade PV systems could be an alternative for many people, especially for those who do not have access to a rooftop. So far, combinations of different vertically tilted modules as well as the combinations between vertical and 45°-tilted surfaces have not yet been taken into account because of the problems with the standardization of shadows from nearby building, trees and, other obstacles.

The calculation in this study assumed a constant price for the FIT over the day. However, if we consider the general trend to link the price of electricity with the spot market price, so that the price of selling or feeding electricity to grid changes according to the production and demand, the E-W and SE-SW orientations might become even more beneficial against S-facing PV systems. In addition, the suitability criteria for rooftops carrying solar modules must be questioned [26]. More roofs should be taken into account when diurnal variations are considered. Based on our measurements and analysis we conclude that the yearly sum of produced electricity can no longer be the only criterion for the installation of PV modules. Instead, other orientations may be more beneficial for both the owner and the society that uses solar power.

Regarding the model validation, both of the tested PV software overestimate the energy production at most studied orientations and also overestimate the percentage of these orientations when compared to the south-oriented generator. This result agrees with previous results [45], which showed that anisotropic models overestimate the S-tilted irradiance and most vertical irradiances. The need to improve existing modelling has also been shown in previous studies [46,47]. A major cause for the deviation between models and measurements may be the oversimplified assumptions about the sky radiance, which can be overcome by new measurement techniques [48,49]. Moreover, the study showed that the overestimation increases with increasing deviation from the south direction. In addition, both PV programs underestimate SC rate and AD for all studied orientations. SC rate was underestimated from 0.4% to 14%, while AD values were underestimated from 1.3% to 8.1%. These results lead to the conclusion that improvements are necessary when modelling SC and AD.

The amount of solar irradiance received by the surface of the PV collector is among the most important parameters that affect the performance of a PV system. Therefore, high-resolution tilted solar irradiance data in various orientations and weather conditions are needed to feed the models for better simulation of PV Power.

Author Contributions: R.M. conceived and designed the study and wrote the draft paper; G.S. initiated the investigation on the performance of differently oriented PV modules. G.S. and E.W.L. contributed in the conception and design, analysis and interpretation of the data. All the authors significantly contributed to the final version of the manuscript.

Funding: The publication of this article was funded by the Open Access fund of Leibniz Universität Hannover" This is what we wrote in all our previous papers.

Acknowledgments: The publication of this article was funded by the Open Access fund of Leibniz Universität Hannover. We are also grateful to Holger Schilke for his contribution in collecting the data and supervising the measurements. Thanks also to Martin Hoffmann for his instruction in using PVSOL. Ben Liley from the National Institute of Water and Atmospheric Research (NIWA), New Zealand for improving the English and providing helpful comments on the clarity of the presentation.

Conflicts of Interest: The authors declare no conflict of interest.

Nomenclature

SC	Self-consumption
AD	Degree of autarky
P_m	Power output of the PV system
P_{rel}	Rated PV system power
I_m	Measured solar irradiance
I_{UTC}	Solar irradiance at STC (1000 W/m^2)
T_{sen}	Sensor temperature
γ	Power temperature coefficient
Pdu	PV directly used energy
PLF	Power loss factor
φ	Geographical latitude φ
PVg	Total PV generated energy
Pe	Electricity price
PG	Grid electricity price
Pfi, FIT	Feed-in tariff
LC	Levelized coast of PV energy

References

1. Jäger-Waldau, A. Snapshot of Photovoltaics 2018. *EPJ Photovolt.* **2018**, *9*, 6. [CrossRef]
2. Ginley, D.; Green, M.A.; Collins, R. Solar energy conversion toward 1 TW. *MRS Bull.* **2008**, *33*, 355–364. [CrossRef]
3. Vartiainen, E.; Breyer, C.; Moser, D.; Medina, E.R. Impact of weighted average cost of capital, capital expenditure, and other parameters on future utility-scale PV levelised cost of electricity. *Prog. Photovolt. Res. Appl.* **2019**, 1–15. [CrossRef]
4. IRENA. Renewable Energy Statistics 2018, The International Renewable Energy Agency. Available online: https://www.irena.org//media/Files/IRENA/Agency/Publication/2019/Mar/RE_capacity_highlights_2019pdf?la=en&hash=BA9D38354390B001DC0CC9BE03EEE559C280013F (accessed on 26 October 2019).
5. Khoo, Y.S.; Nobre, A.; Malhotra, R.; Yang, D.; Ruther, R.; Reindl, T.; Aberle, A.G. Optimal orientation and tilt angle for maximizing in-plane solar irradiation for PV applications in Singapore. *IEEE J. Photovolt.* **2014**, *4*, 647–653. [CrossRef]
6. Hartner, M.; Ortner, A.; Heisl, A.; Haas, R. East to west—The optimal tilt angle and orientation of photovoltaic panels from an electricity system perspective. *Appl. Energy* **2015**, *160*, 94–107. [CrossRef]
7. Seme, S.; Krawczyk, A.; Łada, E.; Tondyra, Š.B.; Hadžiselimović, M. The Efficiency of Different Orientations of Photovoltaic Systems. *Przegląd Elektrotechniczny* **2017**, *93*, 201–204. [CrossRef]
8. Ghazi, S.; Ip, K. The effect of weather conditions on the efficiency of PV panels in southeast of UK. *Renew. Energy* **2014**, *69*, 50–59. [CrossRef]
9. Colli, A.; Zaaiman, W.J. Maximum-power-based PV performance validation method: Application to single-axis tracking and Fixed-Tilt c-Si systems in the Italian Alpine region. *IEEE J. Photovolt* **2012**, *2*, 555–563. [CrossRef]

10. Bakirci, K. General models for optimum tilt angles of solar panels: Turkey case study. *Renew. Sustain. Energy Rev.* **2012**, *16*, 6149–6159. [CrossRef]
11. Yan, R.; Saha, T.K.; Meredith, P.; Goodwin, S. Analysis of yearlong performance of differently tilted photovoltaic systems in Brisbane, Australia. *Energy Convers. Manag.* **2013**, *74*, 102–108. [CrossRef]
12. Lave, M.; Kleissl, J. Optimum fixed orientations and benefits of tracking for capturing solar radiation in the continental United States. *Renew. Energy* **2011**, *36*, 1145–1152. [CrossRef]
13. Agarwal, A.; Vashishtha, V.K.; Mishra, S.N. Comparative approach for the optimization of tilt angle to receive maximum radiation. *Int. J. Eng. Res. Technol.* **2012**, *1*, 1–9.
14. Tang, R.; Wu, T. Optim. tilt-angles for solar collectors used in China. *Appl. Energy* **2004**, *79*, 239–248. [CrossRef]
15. Uba, F.A.; Sarsah, E.A. Optimization of tilt angle for solar collectors in WA, Ghana, Pelagia Research Library. *Adv. Appl. Sci. Res.* **2013**, *4*, 108–114.
16. Huld, T.; Šúri, M.; Dunlop, E.D. Comparison of potential solar electricity output from fixed-inclined and two-axis tracking photovoltaic modules in Europe. *Prog. Photovolt. Res. Appl.* **2008**, *16*, 47–59, ISSN 1099-159X. [CrossRef]
17. Huld, T.; Müller, R.; Gambardella, A. A new solar radiation database for estimating PV performance in Europe and Africa. *Sol. Energy* **2012**, *86*, 1803–1815. [CrossRef]
18. Quinn, S.W.; Lehman, B.A. Simple formula for the optimum tilt angles of photovoltaic panels. In Proceedings of the IEEE 14th Workshop on Control and Modeling for Power Electronics (COMPEL), Salt Lake City, UT, USA, 23–26 June 2013; pp. 1–8.
19. Beringer, S.; Schilke, H.; Lohse, I.; Seckmeyer, G. Case study showing that the tilt angle of photovoltaic plants is nearly irrelevant. *Sol. Energy* **2011**, *85*, 470–476. [CrossRef]
20. Bodis, K.; Kougias, I.; Jäger-Waldau, A.; Taylor, N.; Szabo, S. A high-resolution geospatial assessment of the rooftop solar photovoltaic potential in the European Union. *Renew. Sustain. Energy Rev.* **2019**, *114*, 109309. [CrossRef]
21. Mertens, K. *Photovoltaics: Fundamentals, Technology and Practice*, 1st ed.; John Wiley & Sons, Ltd. Published: Hoboken, NJ, USA, 2014.
22. Luthander, R. Photovoltaic System Layout for Optimized Self-Consumption. Uppsala Universitet. 2013. Available online: http://www.diva-portal.org/smash/get/diva2:637625/FULLTEXT01.pdf (accessed on 26 November 2019).
23. Beck, T.; Kondziella, H.; Huard, G.; Bruckner, T. Assessing the influence of the temporal resolution of electrical load and PV generation profiles on self-consumption and sizing of PV-battery systems. *Appl. Energy* **2016**, *173*, 331–342. [CrossRef]
24. Luthander, R.; Widen, J.; Nilsson, D.; Palm, J. Photovoltaic self consumprion in buidings: A review. *Appl. Energy* **2015**, *142*, 80–94. [CrossRef]
25. Leicester, P.A.; Rowley, P.N.; Goodier, C.I. Probabilistic analysis of solar photovoltaic self-consumption using Bayesian network models. *IET Renew. Power Gener.* **2016**, *10*, 448–455. [CrossRef]
26. Martín, A.M.; Domínguez, J.; Amador, J. Applying LIDAR datasets and GIS based model to evaluate solar potential over roofs: A review. *AIMS Energy* **2015**, *3*, 326–343. [CrossRef]
27. PVSOL. PV*SOL Online—A Free Tool Forsolar Power (PV) Systems. 2019. Available online: http://pvsol-online.valentin-software.com/#/ (accessed on 6 May 2019).
28. *PVSyst, Version, 6.79*; [*Computer Software*]; University of Geneva: Geneva, Switzerland, 2019.
29. DGS. *EnergyMap.info*; Deutsche Gesellschaft für Sonnenenergie e.V.: Berlin, Germany, 2015.
30. PVGIS. Overview of PVGIS data sources and calculation methods. European Commission. Available online: http://re.jrc.ec.europa.eu/pvg_static/methods.html (accessed on 26 November 2019).
31. Idoko, L.; Anaya-Lara, O.; McDonald, A. Enhancing PV modules efficiency and power output using multi-concept cooling technique. *Energy Rep.* **2018**, *4*, 357–369. [CrossRef]
32. Tjaden, T.; Bergner, J.; Weniger, J.; Quaschning, V. Repräsentative elektrische Lastprofile für Wohngebäude in Deutschland auf 1-sekündiger Datenbasis. Technical report, Berlin, Germany. Hochschule für Technik und Wirtschaft HTW Berlin. Available online: https://pvspeicher.htw-berlin.de/wp-content/uploads/2017/05/HTW-BERLIN-2015-Repr%C3%A4sentative-elektrische-Lastprofile-f%C3%BCr-Wohngeb%C3%A4ude-in-Deutschland-auf-1-sek%C3%BCndiger-Datenbasis.pdf (accessed on 26 November 2019).

33. Haysom, J.E.; Hinzer, K.; Wright, D. Impact of electricity tariffs on optimal orientation of photovoltaic modules. *Prog. Photovolt. Res. Appl.* **2015**, *24*, 253–260. [CrossRef]
34. Jager-Waldau, A.; Bucher, C.; Frederiksen, K.H.B.; Guerro-Lemus, R.; Mason, G.; Mather, B.; Mayr, C.; Moneta, D.; Nikoletatos, J.; Roberts, M.B. Self-consumption of electricity produced from PV systems in apartment buildings—Comparison of the situation in Australia, Austria, Denmark, Germany, Greece, Italy, Spain, Switzerland and the USA. In Proceedings of the 2018 IEEE 7th World Conf. Photovolt. Energy Conversion, WCPEC 2018—A Jt. Conf. 45th IEEE PVSC, 28th PVSEC 34th EU PVSEC 1424–1430, Waikoloa Village, HI, USA, 10–15 June 2018. [CrossRef]
35. Mojonero, D.H.; Villacorta, A.R.; Kuong, J.L. Impact assessment of net metering for residential photovoltaic distributed generation in Peru. *Int. J. Renew. Energy Res.* **2018**, *8*, 1200–1207.
36. Pereira da Silva, P.; Dantas, G.; Pereira, G.I.; Câmara, L.; De Castro, N.J. Photovoltaic distributed generation—An international review on diffusion, support policies, and electricity sector regulatory adaptation. *Renew. Sustain. Energy Rev.* **2019**, *103*, 30–39. [CrossRef]
37. BDEW. BDEW-Strompreisanalyse (Haushalte und Industrie). Available online: https://www.bdew.de/media/documents/190115_BDEW-Strompreisanalyse_Januar-2019.pdf (accessed on 28 January 2019).
38. Fraunhofer Institute for Solar Energy Systems (ISE). *Levelized Cost of Electricity Renewable Energy Technologies*; Fraunhofer Institute for Solar Energy Systems (ISE): Freiburg, Germany, 2018.
39. Stenzel, P.; Linssen, J.; Fleer, J.; Busch, F. Impact of temporal resolution of supply and demand profiles on the design of photovoltaic battery systems for increased self-consumption. In Proceedings of the 2016 IEEE International Energy Conference (ENERGYCON), Leuven, Belgium, 4–8 April 2016.
40. Talavera, D.L.; Nofuentes, G.; Aguilera, J. The Internal Rate of Return of Photovoltaic Grid-Connected Systems: A Comprehensive Sensitivity. *Renew. Energy* **2010**, *35*, 101–111. [CrossRef]
41. Tapia, M. Evaluation of Performance Models against Actual Performance of Grid Connected PV Systems. Master's Thesis, Carl von Ossietzky Universität, Oldenburg, Germany, 2014.
42. Fraunhofer ISE. *Recent Facts about Photovoltaics in Germany*; Fraunhofer Institute for Solar Energy Systems ISE: Freiburg, Gemmary, 2019.
43. Obane, H.; Okajima, K. Extracting issues significant to valuing electricity from small photovoltaic systems using quantitative content analysis. *Electr. J.* **2019**, *32*, 106673. [CrossRef]
44. Solar Power Europe. *Global Market Outlook for Solar Power 2018–2022*; Solar Power Europe: Brussels, Belgium, 2018.
45. Mubarak, R.; Hofmann, M.; Riechelmann, S.; Seckmeyer, G. Comparison of modelled and measured tilted solar irradiance for photovoltaic applications. *Energies* **2017**, *10*, 1688. [CrossRef]
46. Hofmann, M.; Seckmeyer, G. Influence of various irradiance models and their combination on simulation results of photovoltaic systems. *Energies* **2017**, *10*, 1495. [CrossRef]
47. Hofmann, M.; Seckmeyer, G. A New Model for Estimating the Diffuse Fraction of Solar Irradiance for PV System Simulations. *Energies* **2017**, *10*, 248. [CrossRef]
48. Seckmeyer, G.; Lagos Rivas, L.; Gaetani, C.; Heinzel, J.W.; Schrempf, M. Biologische und medizinische Wirkungen solarer Strahlung (Biological and medical effects of solar radiation). In *Promet, Heft 100 Strahlungsbilanzen*; Chapter 13; Deutscher Wetterdienst (DWD): Offenbach am Main, Germany, 2018.
49. Riechelmann, S.; Schrempf, M.; Seckmeyer, G. Simultaneous measurement of spectral sky radiance by a non-scanning multidirectional spectroradiometer (MUDIS). *Meas. Sci. Technol.* **2013**, *24*, 125501. [CrossRef]

Review

Grid Synchronization and Islanding Detection Methods for Single-Stage Photovoltaic Systems

Rosa Anna Mastromauro

Department of Information Engineering (DINFO), University of Florence, 50139 Florence, Italy; rosaanna.mastromauro@unifi.it; Tel.: +39-055-275-8650

Received: 29 April 2020; Accepted: 28 June 2020; Published: 1 July 2020

Abstract: Synchronization and islanding detection represent some of the main issues for grid-connected photovoltaic systems (PVSs). The synchronization technique allows to achieve PVS high power factor operation and it provides grid voltage monitoring. The islanding detection control function ensures safe operation of the PVS. Focusing on low-power single-stage PVSs, in this study the most adopted and the highest performance synchronization and islanding detection methods are discussed. The role of the synchronization system is fundamental to detect the grid conditions, for the islanding detection purpose, and to manage the reconnection to the grid after a PVS trip. Hence a combined review is advantageous.

Keywords: photovoltaic systems; synchronization systems; phase-locked loops; islanding detection methods

---

1. Introduction

At the end of 2018, the world had 152 GW of installed photovoltaic (PV) electricity capacity. The best PV markets in 2018 were China with 44.3 GW, India with 10.8 GW, USA with 10.7 GW, Japan with 6.7 GW, Australia with 3.8 GW. The European Union (EU) has registered rise for the first time in years with 8.4 GW, but this growth is far from the 23.2 GW registered in 2011. However, the growth can be considered slow, some of the EU countries have already achieved high PV penetration due to past installations such as Germany with a PV overall capacity of 45.5 GW by the end of 2018, Italy that exceeds 20 GW, United Kingdom with 13 GW, Spain with 5.6 GW, Belgium with 4.3 GW and Switzerland with 2.2 GW. It is estimated that overall, the PV systems (PVSs) had contributed to the 2.9% of the global electricity demand in 2018 and that the climate change impact is of 590 millions of tons of CO_2 saving every year [1–3].

Due to the photovoltaic prize reduction and availability of loan products, a significant portion of PVSs have been recently installed also in absence of governments initiatives especially for residential applications. A performance evaluation of residential PVSs in some European countries is presented in [4]. Considering the period 2014–2016, the highest specific yield in kWh/kWp has been registered in Italy in 2015.

The PVSs inverters price has diminished around 0.10 $/Wp in the last decade [5]. In addition, the design optimization of the PVS converters has facilitated the reduction of the total cost of ownership [6]. Nevertheless, the increase of PVS grid-connected installations implies several management challenges depending also on the point of interconnection between the PVS and the grid [7–10]. In this scenario advanced control features of the PVS inverters can contribute to overcome some of the grid management challenges due to high penetration [11–13].

Looking at the residential applications, the PVSs can be single-stage or double-stage. In case of single-stage PVSs, the PV array is directly connected to the inverter avoiding a boost DC/DC converter. Single-stage transformerless PVSs represent the most promising technology due to lower weight, higher efficiency, smaller size and limited cost than double-stage PVSs or single-stage architectures coupled to low-frequency transformers [14–18]. Focusing on a single-stage PVS, a review about some of the main control issues is presented in [19], however the analysis is limited to current and voltage control methods and maximum power point tracking (MPPT) techniques.

About the most important issues to be considered in grid-connected PVS there are the synchronization with the grid and the detection of the islanding condition. Synchronization deals with PVS high power factor operation, since the synchronization algorithms objective is to provide grid voltage information about amplitude, phase and frequency in order to generate a current/voltage reference which is in phase with the grid voltage [13,20–24].

Synchronization deals also with the grid voltage monitoring. According to the grid-connection requirements [25], the PVSs connected to the low-voltage distribution grid must operate without causing step change in the RMS voltage at the point of common coupling (PCC) exceeding 5% of rated value. In addition, the synchronization parameters limits for grid-connected PVS are: 0.3 Hz for the frequency difference, 10% for the voltage difference. Abnormal conditions can arise on the utility grid which require a prompt response from the grid-connected PVS, hence the information provided by the synchronization system are fundamental for this purpose [26–28].

Unintentional islanding phenomenon is verified in case of grid power outages when the PVS continues to supply the local loads. Unintentional islanding can cause damages to the local electrical loads, to the grid-connected PVS inverter, to the technicians during the maintenance operations. Numerous improved islanding detection algorithms have been proposed in literature in the last years aiming to detect islanding phenomenon in all possible cases [29–35]. However, many these algorithms are not designed peculiarly for PVS.

In case of low power residential PVSs and in particular in case of single-stage systems, the PVS inverter is commonly in charge of the islanding detection, hence the anti-islanding functionality represents one of the main challenge in the PVS inverters design [18]. The anti-islanding protections must be implemented on the basis of the international standards requirements for distributed power generation systems (DPGSs) [25,36–38]. In particular, it is required that unintentional islanding be detected in less than two seconds as already established in the previous guidelines for PVSs [39–41].

After a disconnection due to the islanding detection an improper reconnection event is not improbable if the PVS breaker connects the system to the grid when the PVS voltage is out of phase. In this hypothesis a second disconnection can occur due to the PVS protections action. Hence the reclosing procedure has to be managed in strict coordination with the PVS synchronization system. For this reason, synchronization and islanding detection issues must be analyzed together.

About synchronization systems some books have been published such as [42]. Few review papers can be found in literature [23,43–46]. In [43] all the main families of synchronization techniques (also including the artificial intelligence techniques) are classified showing advantages and disadvantages. Basic concepts about phase-locked loop (PLL) techniques are explained in [44]. Reference [45] is devoted to three-phase applications, reference [23] is devoted to single-phase application, while [46] is oriented to design issues.

About islanding detection methods many review papers have been published in literature considering different DPGSs [47–57]. Reference [47] provides a review of the islanding detection methods for high power DPGSs. In [48] an extensive review of the islanding detection methods is provided focusing on some performance indices evaluation and in particular on the detection time. Reference [49] is focused just on passive methods, reference [50] is focused just on active methods. In [51] the focus is on active and passive methods and a new active methods is proposed for a three-phase PVS. The same islanding detection methods are discussed also in [52] also including the hybrid detection methods. However, hybrid methods are categorized just as combination of active and passive methods. In [53] some active islanding detection techniques are compared on the basis of a new index assessing the non-detection-zone (NDZ) size. In [54] the active techniques are classified in two categories: techniques introducing positive feedback in the control of the inverter and techniques based on harmonics injection. In [55] the islanding detection methods based on different signal processing techniques are discussed in detail. Reference [56] provides a comprehensive review of the islanding detection techniques particularly oriented to recent intelligence-based methods. A similar approach is adopted in [57]. Intelligence-based islanding detection is out of topic in relation to the present study.

All the analyzed papers share as starting point a first classification among remote islanding detection techniques (based on communication) and local islanding detection techniques. Currently this classification can be overtaken considering the availability of communication protocols and local communication interface equipment by most PVS inverters.

No studies combining the analysis of the synchronization systems with the islanding detection techniques can be found in literature for PVSs or for other DPGSs applications. Nevertheless, many islanding detection schemes are based on modification or employment of additional synchronization systems to assess the unintentional islanding condition. In addition, independently of the adopted islanding detection technique, two synchronization systems are required in order to manage the reconnection procedure to the main grid by a PVS after an islanding event. Starting from this issues, the aim of this study is to provide a combined analysis about synchronization and islanding detection techniques which need coordinated operation and proper integration in the PVSs control structure.

Focusing on single-stage single-phase PVSs, the present study aims at giving an update of the most-recent trends about synchronization techniques and islanding detection methods, in particular: in Section 2 there are summarized the main goals of the single-stage PVSs control systems. The most adopted and the highest performance grid synchronization methods are analyzed in Section 3; while Section 4 is an overview of the islanding detection methods. Section 5 is about coordination between synchronization and islanding detection systems. Finally, the conclusions are presented in Section 6.

2. Control System Functionalities of a Single-Stage photovoltaic Power System

The overall control structure of a single-stage PVS is shown in Figure 1 where it is assumed that the PVS can be connected to a local load, to the utility grid or it can be part of a smartgrid. The control functionalities can be classified in basic control functions and ancillary control functions. The basic control functions are the maximum power extraction, the grid synchronization, the current and the voltage control, the unintentional islanding detection. High power factor operation and harmonic rejection are achieved by proper design of current and voltage controllers and of the synchronization system. The current/voltage control reference signal is provided by the PV source power control which consists of a maximum power point tracking (MPPT) algorithm and a DC voltage controller. The MPPT algorithm is in charge of the maximum power extraction.

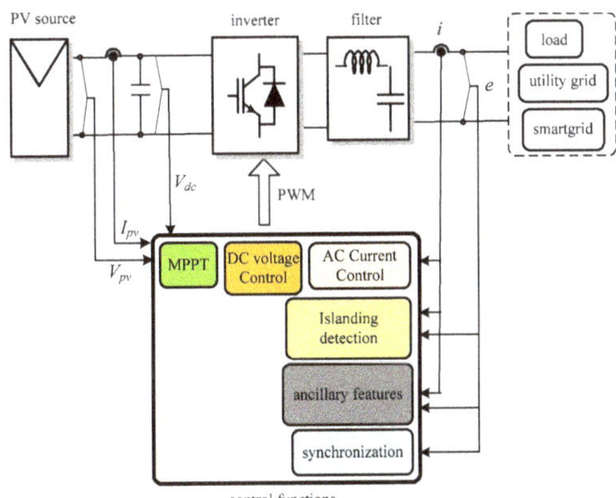

Figure 1. Single-stage photovoltaic systems (PVSs) control functions.

The ancillary control functions are the ride-through capability, the voltage and the frequency support to the local loads or the main grid. The ancillary control functions are out of the scope of the present study.

3. Grid Synchronization

The phase angle of the grid voltage is a critical piece of information for grid-connected systems since it is used to obtain the control reference signal as previously pointed out. Numerous methods using different techniques for synchronization and grid-voltage monitoring have been presented in the technical literature about DPGS. Most of these studies are related to three-phase systems [22,28,58,59] than to single-phase applications [60,61]. Some of the methods are not always categorized properly, thus leading to confusion. In order to clarify, the most used techniques can be organized as presented in Figure 2.

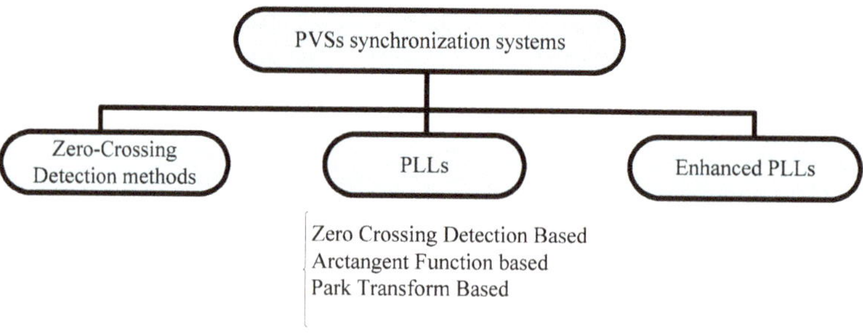

Figure 2. PVSs synchronization techniques classification.

Commonly the synchronization and grid voltage monitoring methods are classified in two main categories: zero-crossing detection (ZCD) methods [62] avoiding the grid voltage phase control, phase-locked loop (PLL) methods based on a strict control of the grid voltage phase [63]. The PLL techniques are generally categorized as PLL based on ZCD, on the arctangent function [64] and on the Park transform [44]. Among these, the Park transform-based PLL provides the best performance. Actually, in the last years, a new family of PLLs known as enhanced PLL has been ranked as the most promising.

Looking at the single-stage PVS shown in Figure 1, two synchronization systems are required in order to manage correctly the disconnections and reconnections with the main power system: the first synchronization system is used to monitor the voltage grid, the second one is used to monitor the PVS voltage. Among the synchronization systems presented in literature, only some methods are compatible with the considered application.

3.1. Zero-Crossing Detection Methods

An elementary method used to extract information about phase and frequency of the grid voltage is based on the zero-crossing measurement [62,65–67]. The ZCD structure is shown in Figure 3. When the grid voltage waveform crosses the zero, a counter provides in output information about the period and, consequently, the estimated frequency $\hat{\omega}$ of the grid voltage is obtained. The phase $\hat{\theta}$ of the grid voltage is achieved integrating $\hat{\omega}$. Despite the simplicity, this technique does not allow high dynamic performances. Indeed, the phase tracking can be fulfilled just for each half cycle of the grid voltage waveform.

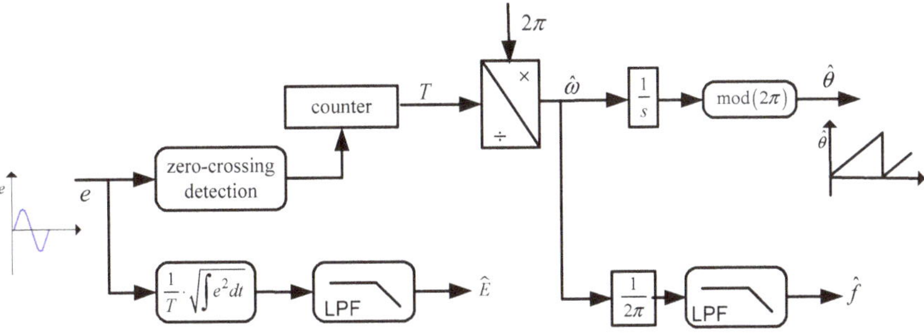

Figure 3. Zero-crossing detection (ZCD) structure.

Since the grid voltage e is generally affected by power quality disturbances, a low-pass filter (LPF) is used to extract the fundamental frequency f of the original signal. No controller is employed in the detection and the method is not proper to track and to monitor the grid voltage in case of abrupt changes.

3.2. PLLs

The PLL consists of a phase comparator and a PI controller. The phase comparator determines the phase error ε which is provided in input to the PI. A reference frequency ω_{IC} and the output of the PI are summed in order to evaluate the grid voltage frequency $\hat{\omega}$. The feed-forward action allows to improve the PLL dynamic performance. Later the grid voltage angle $\hat{\theta}$ is calculated by the information of the grid voltage frequency. The phase comparator operation can be based on a reference signal provided by the ZCD of the input grid voltage, by the arctangent function of by the Park transform.

In case of ZCD PLL, the ZCD discussed in the previous subsection is employed to extract the phase reference for the phase comparator [42,68]. As described before, also this synchronization system is not proper to track the grid voltage in case of abrupt variations. However, the ZCD PLL provides better performance than the ZCD technique since the estimated angle $\hat{\theta}$ is controlled in closed loop.

Both the arctangent function-based PLLs and the Park transform-based PLLs require a voltage orthogonal system. In case of the arctangent function-based PLL [42,64], the phase reference for the phase comparator is extracted calculating the arctangent by e_α and e_β information. The disadvantage is that the arctangent function is not easy to implement.

In Figure 4 there is shown the structure of a PLL-based on the Park transform which represents the most adopted solution.

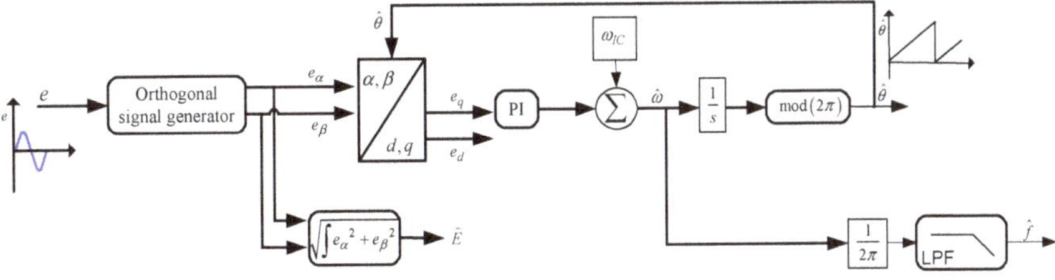

Figure 4. Phase-locked loop (PLL) structure based on the Park transform.

A coordinate transformation from $\alpha\beta$ to dq is usually adopted to process DC signals instead of AC signals. The grid voltage phase angle is extracted synchronizing the grid voltage vector with the dq rotating reference frame. Forcing the q-axis voltage reference to zero, the lock with the grid voltage is ensured [69]. The error signal is processed by a PI. The output of the PI controller is the grid frequency

The estimated frequency is integrated; hence the grid voltage phase angle is measured, and the result is provided in input to the $\alpha\beta$-dq transformation block. As shown in Figure 4, the phase detection is based on the Park transform [70] while the PI controller acts as a filter which determines the dynamics of the phase lock. For this reason, the PI controller parameters are chosen considering the tradeoff between filtering performance and fast dynamics [46].

All the PLLs described up to now need a LPF as in case of the ZCD synchronization method. It occurs to extract the fundamental frequency $\hat{f}$ of the original signal e which is commonly affected by harmonic disturbances. Neglecting the LPF, the detailed structure of the PLL based on the Park transform is depicted in Figure 5 for a better understanding of how the phase comparator is obtained through the use of the Park transform.

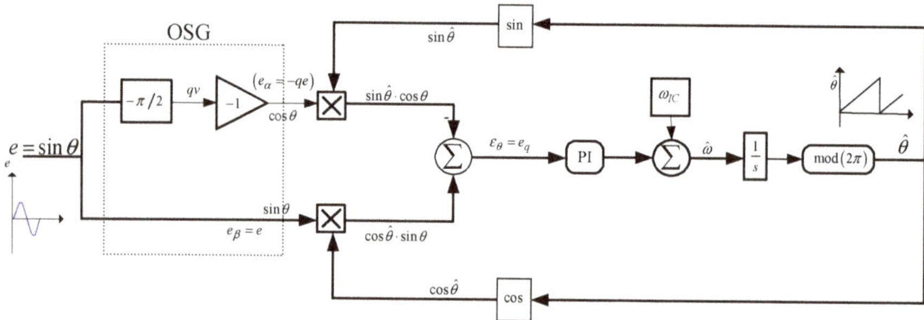

Figure 5. Detailed PLL structure based on the Park transform.

In case of single-phase systems, the orthogonal voltage system has to be artificially generated [21,71] and it represents the main challenge for the grid voltage monitoring. The orthogonal signal generator (OSG) is in charge of the orthogonal voltage system realization. One of the most advanced technique adopts the second order generalized integrator (SOGI) [72,73]. The OSG based on the SOGI filter allows also to extract the fundamental component of the grid voltage, for this reason the LPF used to extract $\hat{f}$ can be avoided in case of the SOGI PLL. The OSG based on the SOGI filter is shown in Figure 6.

The grid voltage e is transformed in two sinusoidal signals denoted as e' and qe'. e' and qe are phase shifted of $\pi/2$. The sinusoidal signal e' is in phase with the grid voltage e. In addition, e' and the first harmonic component of the grid voltage exhibit the same magnitude.

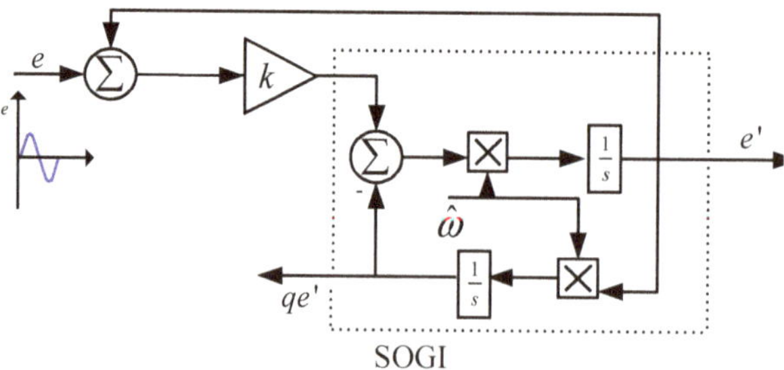

Figure 6. Orthogonal signal generator (OSG)–second-order generalized integrator (SOGI) standard structure.

The SOGI acts like an infinite gain band-pass filter whose transfer function is defined as:

$$H_{SOGI}(s) = \frac{\omega_n s}{s^2 + \omega_n^2} \tag{1}$$

where ω_n represents the undamped natural frequency of the SOGI which should coincide with the estimated frequency ($\omega_n = \hat{\omega}$).

The performances of all the single-phase PLLs based on a OSG are particularly affected by the voltage offset commonly introduced by the measurement equipment and by the signal processing operation [27]. The frequency of the error derived by the grid voltage offset is the same of the grid voltage waveform. The PLLs based on a OSG are not properly designed to provide rejection to the voltage offset. However, since the PI controller acts as a filter, the PLL controller parameters could be tuned in order to achieve filtering of the voltage offset. It would modify the bandwidth of the overall system, but, unfortunately, it would impact considerably the dynamic performances of the PLL. In [71] the performances of single-phase PLLs based on different OSGs are compared and a guideline for the PLLs parameters tuning is proposed.

3.3. EPLLs

In the last years, an alternative synchronization technique known as enhanced phase-locked loop (EPLL) has succeeded [23,74] due to high filtering performance. It consists of an adaptive nonlinear detection algorithm which provides two orthogonal signals synchronized with the grid voltage. The EPLL structure is represented in Figure 7. It allows to estimate the frequency, the phase and also the amplitude E_{EPLL} of the input signal fundamental component. The EPLL operates as an adaptive filter (either a notch or a band–pass filter) whose frequency tracks the fundamental frequency of the grid voltage. e_{EPLL} denotes the filtered signal tracking the grid voltage supplied in input.

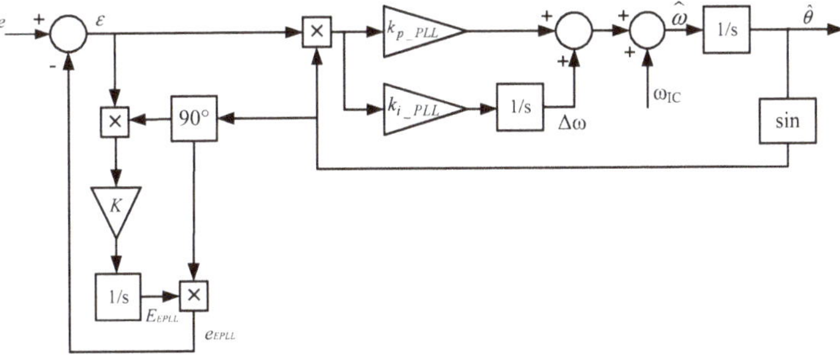

Figure 7. Enhanced phase-locked loop (EPLL) structure.

The EPLL represents one of the most promising synchronization systems for single-phase applications since it provides: filtering capability in respect to the undesired harmonics, adaptive detection of the grid voltage fundamental frequency, proper estimation of frequency, angle and amplitude of the grid voltage supplied in input [24,75]. Some modifications of the EPLL have been also proposed in the most recent literature, in particular in [76] the structure of the EPLL has been modified in order to achieve a linear model.

The main differences of the EPLL compared to the PLL based on the Park transform occur testing the two systems in presence of grid voltage perturbations, in particular frequency changes and harmonics. It has been demonstrated that the EPLL exhibits higher filtering capability and shorter transients [23,77]. For all these reasons the EPLL represents the ideal candidate to operate in coordination with the islanding detection techniques discussed in the following Section.

4. Islanding Detection

In case of grid disconnection, the PVS operation depends on the power level provided by the PVS before islanding occurrence. In Figure 8 it is represented the PVS power stage for the islanding detection test. The grid utility breaker is denoted as Sg, two different breakers S_1 and S_2 are used to connect the PVS to the point of common coupling (PCC) and to connect the load. A variable RLC load is considered in order to assess the islanding phenomenon in case of different load powers and quality factors.

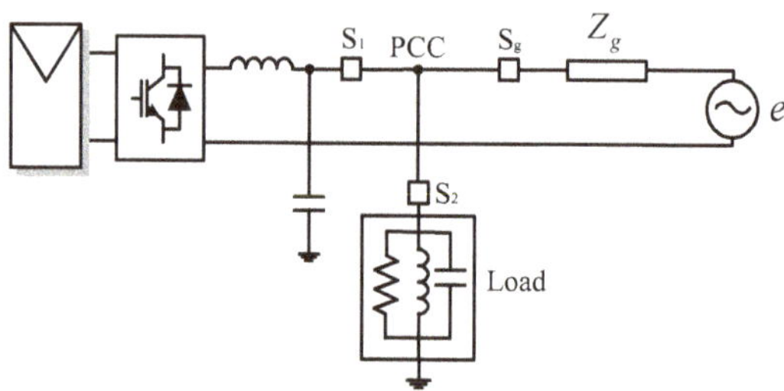

Figure 8. Photovoltaic system (PVS) power stage for islanding detection test.

For each islanding detection method, the non-detection-zone (NDZ) defines the area where the anti-islanding methods fail to detect islanding. As a consequence, the NDZ can be used as a performance index to assess the islanding detection methods [78–80]. However, in comparison with the previous version of the standard [81], the new standard [25] requires voltage and frequency ride-through capability of the PVSs which increases the NDZ of the islanding detection algorithms. Hence it has to be pointed out that ride-through requirements hazard the islanding detection techniques.

Traditionally the islanding detection methods were classified in remote techniques (based on communication signals) and local techniques. Considering the recent advancements of communication equipment and the requirements updates related to the PVS standards, in this study a different classification is adopted. The islanding detection methods are classified in four main categories: Communication-based methods [82], passive methods [49], active methods [25,50] and signal processing-based methods [55]. In addition, in order to improve the performance of the islanding detection methods and to satisfy the standard requirements, hybrid techniques are developing in the last years which are based on combination of the previous categories. In Figure 9 the main islanding detection methods are summarized.

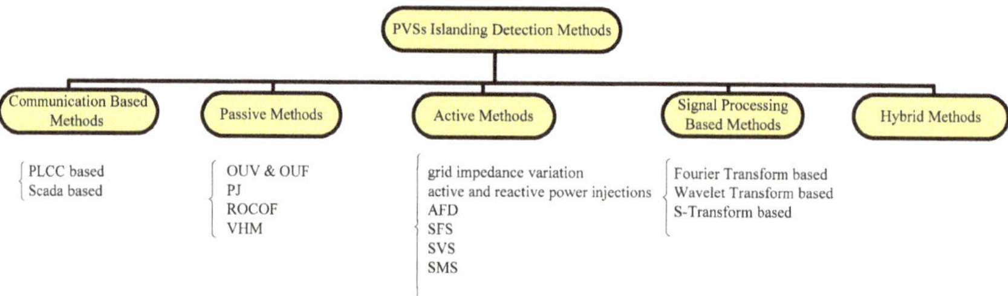

Figure 9. PVSs islanding detection methods classification.

4.1. Communication-Based Methods

Communication-based methods allow to achieve accurate and reliable assessment of the islanding conditions. Otherwise, considering the number of PVSs to be managed and the power size, the required equipment can vary [82,83]. Hence these solutions are not common in case of low voltage and low power residential PVSs. The communication-based methods can be classified in: power line carrier communication (PLCC) methods, Supervisory Control Moreover, data acquisition (SCADA)-based methods.

4.1.1. PLCC-Based Methods

The PLCC-based methods allow a continuous test of the grid connection. A continuous carrier is adopted [84–87] and all the PVSs are equipped with a receiver. When the grid breaker Sg is opened as a consequence of grid disconnection, the receivers do not catch signals and the islanding condition is detected. The PLCC-based techniques are particularly resilient to noise since the PVSs inverters switching frequencies do not influence the PLCC signals.

4.1.2. SCADA-Based Methods

The SCADA-based methods assess the status of the breakers connecting the PVSs to the grid [57,88,89]. Using the SCADA methods, the coordination between the PVS and the grid utility operations is significantly improved, and the islanding phenomenon can be detected avoiding a NDZ area. Cost and complexity of the equipment represent the main disadvantages.

4.2. Passive Methods

Compared to the communication-based methods, the passive islanding detection methods present simple implementation based on protection relays and synchronization systems [41,49]. The most famous passive methods are over/under voltage (OUV) and over-under frequency (OUF), phase jump (PJ), rate of change of frequency (ROCOF), voltage harmonic monitoring (VHM).

4.2.1. OUV and OUF Methods

All grid-connected PVS inverters are required to have OUV and OUF protections. The aim is to avoid power supply by the PVSs when the voltage amplitude and frequency values at the PCC are different from set values [41,90].

Considering an RLC load whose resonant frequency is equal to the grid frequency, no reactive power absorption is verified by the load. In case of grid disconnection, the power absorbed by the load is equal to the active power provided by the PVS. Hence the RMS value of the voltage provided by the PVS at the PCC changes from $E_{PVS} = E$ before the disconnection to:

$$E_{PVS} = \delta E \tag{2}$$

where E is the rated RMS value of the grid voltage,

$$\delta = \sqrt{\frac{P_{PVS}}{P_L}} \tag{3}$$

P_{PVS} is the active power supplied by the PVS, P_L is the rated load active power.

In conclusion the voltage value at the PCC increases or decreases depending on the PVS power generation. As a consequence, also the reactive power changes on the basis of the following relationship:

$$Q_{PVS} = \left(\left(\frac{1}{L\omega_{PVS}} \right) - C\omega_{PVS} \right) E_{PVS}^2 \tag{4}$$

In (4) L and C are the inductive and capacitive components of the RLC load, ω_{PVS} denotes the voltage frequency at the PCC after the grid disconnection. Hence it results:

$$\omega_{PVS} = \frac{-\left(\frac{Q_{PVS}}{E_{PVS}^2 C} \right) + \sqrt{\left(\frac{Q_{PVS}}{E_{PVS}^2 C} \right)^2 + \frac{4}{LC}}}{2} \tag{5}$$

The voltage frequency and amplitude variations allow to detect islanding operation. Unfortunately, in case of power balance between the PVS generation and the load, no active and reactive power variations are registered and consequently no voltage frequency and amplitude variations can be measured. Similarly, small active and reactive power variations imply small voltage variations in terms of frequency and amplitude. For this reason, the OUV and OUF protection cannot detect islanding. OUV and OUF protections are considered insufficient anti-islanding techniques since the active and reactive power variations due to the islanding phenomenon are commonly limited and, as a consequence, there is high probability to fall into the NDZ.

4.2.2. Phase Jump Method

The aim of the phase jump (PJ) method is the detection of a "jump" between the PVS inverter current and voltage [78]. The PJ technique represents one of the first anti-islanding methods and it is based on the use of synchronization systems extracting information about the current and the voltage phase. However, considering the present availability of fast PLLs and the development of high performance current controllers, voltage and current synchronization could be achieved also during islanding operation. Consequently, the detection of the islanding phenomenon based on this technique could fail.

4.2.3. Rate of Change of Frequency Method

The ROCOF islanding detection method measures the rate of change of frequency df/dt in a set time window. When the grid is disconnected, the power mismatch between generation and load causes frequency variations. The PVS is tripped when df/dt exceeds the threshold value [91,92]. The threshold setting is the main issue of this method since it is necessary to distinguish islanding from load changes. Besides the ROCOF exhibits a wide NDZ combined with slow dynamics.

4.2.4. Voltage Harmonic Monitoring Method

The voltage harmonic monitoring (VHM) method is based on the voltage harmonic distortion estimation to detect the occurrence of the islanding phenomenon [49]. In grid-connected operation the voltage at the PCC is set by the grid, but, in case of grid disconnection, the PVS inverter determines the voltage at the PCC. Nevertheless, the voltage harmonic distortion varies with the grid impedance and it depends on the loads connected to the PCC. As a consequence, the accuracy of the method can be hazarded if the islanding detection thresholds are not properly set. Better performance can be achieved monitoring some selected harmonics variations rather than the overall voltage harmonic distortion. In this hypothesis the harmonics variations can be detected by means of PLLs tuned in order to track the selected harmonic components.

4.3. Active Methods

The active islanding detection methods are developed with the goal to achieve better performance than the passive methods. The active methods introduce a perturbation in the PVS through the injection of an active signal [50,93]. The active signal injection is designed considering starting from ideal operating conditions of the PVS. The main active methods can be classified in: Grid impedance variation methods, active and reactive power injections methods, active frequency drift (AFD), Sandia frequency shift (SFS), Sandia voltage shift (SVS), slip-mode frequency shift (SMS).

4.3.1. Grid Impedance Variation Methods

Islanding phenomenon assessment can be based on grid impedance variations monitoring [78,94,95]. A small harmonic current component is drained into the PVS. The grid impedance is evaluated at the frequency of the injected harmonic component. Additional equipment can be employed to measure the grid impedance. Otherwise the grid impedance measurement can be embedded in the PVS inverter control system.

Stability and power quality issues must be tackled when this technique is applied to numerous PVS connected in parallel due to the combination of the injected perturbations and possible inter-harmonics.

4.3.2. Active and Reactive Power Injections Methods

The rationale of the active power injections method is to use controlled active power injections causing active power variations ΔP_{PVS} in the PVS. Consequently, voltage variations can be observed exceeding the threshold voltage value of the islanding protections [50]. Assuming a resistive load R, whose power P_L is constant, it is possible to express the power provided by the PVS as function of the voltage at the PCC. In case of islanding condition, it results:

$$P_{PVS} = P_L = \frac{E_{PVS}^2}{R} \tag{6}$$

Hence it can be obtained:

$$\frac{\partial P_{PVS}}{\partial E_{PVS}} = 2 \cdot \frac{E_{PVS}}{R} = 2 \cdot \frac{\sqrt{R \cdot P_{PVS}}}{R} = 2 \cdot \sqrt{\frac{P_{PVS}}{R}} \tag{7}$$

The voltage variation can be evaluated as:

$$\Delta E_{PVS} = \frac{\Delta P_{PVS}}{2} \cdot \sqrt{\frac{R}{P_{PVS}}} \tag{8}$$

The method is effective but is requires some tuning procedure in order to avoid overcurrents due to the active power injections. The disadvantage is that the islanding technique has to be coordinated with the MPPT operation. Hence the main challenge is to determine when the active power injection can be applied without jeopardizing the other control functions.

Similarly, reactive power injections can be used to cause reactive power variations ΔQ_{PVS} in the system [96]. As a consequence, frequency variations are obtained exceeding the frequency threshold value and islanding condition can be detected.

4.3.3. Active Frequency Drift

The active frequency drift (AFD) is based on a perturbation of the PVS inverter current. In particular, the PVS inverter current reference is modified adding a disturbance current [29]. Denoting as T the period of the grid voltage, it occurs that the PVS inverter current is null for a time portion indicated with t_c in each half cycle as shown in Figure 10.

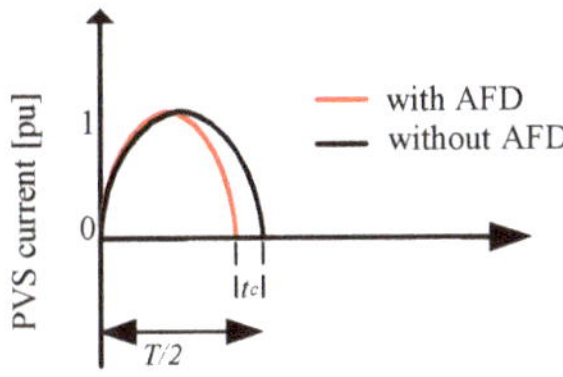

Figure 10. Half wave of the PVS inverter current with and without active frequency drift (AFD).

During grid-connected operation the PVS inverter voltage is not affected by the perturbation. Differently in islanding operation the PVS inverter voltage "drifts" up or down as a consequence of the continuous inverter current variation. The frequency change has to be detected by the UOF protections. Successively the PVS is promptly disconnected.

The AFD method is characterized by a chopping factor c which measures the amount of the perturbation:

$$c = \frac{2t_c}{T} \tag{9}$$

The method ensures correct islanding detection and simple implementation, but the main disadvantage is the power quality detriment due to the current variation. Hence the method is not appropriate in case of numerous PVSs connected in parallel.

4.3.4. Sandia Frequency Shift

The Sandia frequency shift (SFS) methods derives from the AFD technique. SFS perturbs the PVS adding a dead-time to the PVS inverter reference current. Hence the inverter current exhibits a phase "shift" [29,97–99].

For this method, the chopping factor c_S is defined as:

$$c_S = c_{S_0} + k_{SFS}(f_{PVS} - f) \tag{10}$$

where k_{SFS} is a proportional gain, c_{S_0} is the chopping factor in absence of frequency error, f is the grid frequency and f_{PVS} is the output frequency of the PVS inverter.

When the PVS is connected to the grid, the frequency error is null since the grid sets the frequency at the PCC. On the contrary, during islanding operation, the frequency error is not negligible. As a consequence, the PVS inverter current grows in order to overcome the phase shift, the chopping frequency increases, and the frequency varies beyond the OUF threshold values. Hence islanding is detected.

The SFS NDZ is smaller than the AFD NDZ. The improved islanding detection performances are achieved to the detriment of the power quality performances. In addition, in case of high PVSs penetration, unexpected transient disturbances can be registered.

4.3.5. Sandia Voltage Shift

The Sandia voltage shift (SVS) method operates with positive feedback of the PCC voltage amplitude [29,100]. During grid-connected operation of the PVS no significant variations are observed. In case of disconnection of the grid, voltage variations are monitored at the PCC. As a consequence, also the PVS inverter voltage varies and later the PVS is tripped since the voltage variations exceed the OUV protections threshold values. The NDZ of the SVS method is very small. However, the power quality is worsened and also the efficiency is reduced since the power processed by the inverter varies.

4.3.6. Slip-Mode Frequency Shift

The slip-mode frequency shift (SMS) detects islanding phenomenon using positive feedback to lead the PVS towards instability in case of grid disconnection [80,101,102]. In case of grid disconnection, the PVS frequency changes naturally. The PVS PLL action can be modified in order to increase the frequency rate of change rather than to annul it. The phase is forced to be a function of the voltage frequency at the PCC. The PLL acts to increase the frequency until the PVS inverter voltage phase grows faster than the phase of the RLC load (unstable region). The PVS is tripped when the inverter voltage frequency exceeds the threshold value. The method can fail when the load phase slope is higher than the slope achieved by the SMS technique. In this case instability could not be recognized.

In Figure 11 there is shown how the action of the PVS PLL is modified on the basis of the SMS rationale. The PVS phase angle changes from $\hat{\theta}_{PVS}$ to θ_{SMS}.

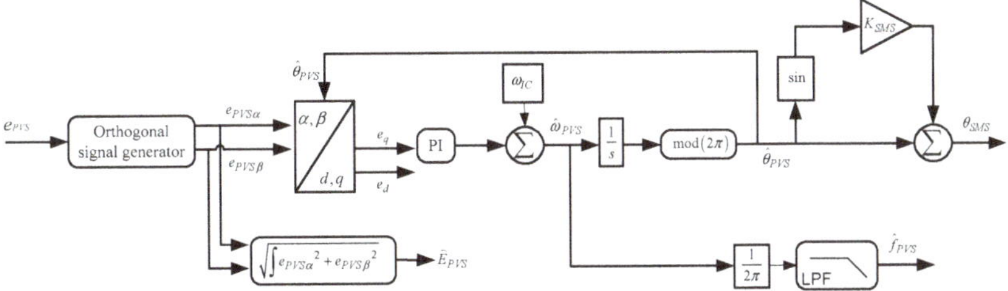

Figure 11. PVS PLL operation in case of SMS islanding-detection technique.

4.4. Signal Processing-Based Methods

Signal processing techniques can be adopted to design new islanding detection algorithms or to improve the performance of the previous developed algorithms [55,82]. The signal processing islanding detection methods are generally based on (a) Fourier transform; (b) wavelet transform and (c) S-transform.

4.4.1. Fourier Transform-Based Methods

The PVS output power is typically affected by variations after grid disconnection, as a consequence, its spectrum varies with continuity in a certain frequency range. Fourier transform (FT) is not proper for the analysis of non-stationary signals. Hence it cannot provide information about fluctuating signals linked to the islanding occurrence [103]. For this reason, the application of this signal processing technique to the analysis of transient phenomena such as islanding is not very common. However, some islanding detection techniques based on the Discrete Fourier transform (DFT) and its modifications have been discussed in literature [104,105]. In [104] a modified VHM islanding detection technique is proposed. Since the equivalent harmonic components measured at the PCC change in case of islanding occurrence, the DFT is employed to assess harmonic components variations. Differently in [105] a kind of FT, named Goertzel algorithm, is employed to develop an active islanding detection method where the Goertzel algorithm extracts the magnitude and phase of some selected components with limited computational burden.

4.4.2. Wavelet Transform-Based Methods

Filters based on the wavelet transform (WT) can track the PVS output power spectrum variations in a certain frequency range. Indeed, WT can process simultaneously signals varying in time and in frequency also in case of non-stationary waveforms. The signal to be processed is decomposed in different levels and the coefficient vectors of each level vary with the signal length. Numerous islanding detection methods based on different mother wavelets were developed [106–111]. As an example, in [108] the multiresolution analysis (MRA) based on the WT is employed to break down the DPGS voltage into different scales. In [109] the WT is adopted just to enhance the performance of the conventional islanding detection methods. In [110] the WT and the back propagation neural network (BPNN) are combined to provide a new islanding detection method based on the normalized logarithmic energy entropy estimation. In [111] the islanding detection technique is based on localization of high-frequency harmonics due to the PVS inverter switching. WT-based algorithms provide generally high performance islanding detection techniques with limited computational burden and implementation complexity. On the contrary these techniques suffer particularly for noise.

4.4.3. S-Transform-Based Methods

S-transform (ST) represents a superior signal processing technique which was born to overcome the noise sensitivity issue of the WT. Various high performance islanding detection techniques based on the ST have been proposed in [112–114]. In [112,113] there are compared the performances of the ST and of the WT. Better results in terms of islanding detection and localization are obtained using the ST instead of the WT. Similar results are achieved in [114] where the islanding detection method is based on the analysis of the negative sequence voltage. For this reason, the method cannot be applied to single-phase PVS avoiding modifications. However, it is demonstrated that the use of the ST results particularly advantageous to detect islanding condition also in presence of noise. In conclusion, higher performances are ensured in the islanding detection since S-transform allows the extraction of the phase of each frequency component related to the time-varying signals involved in the islanding phenomenon. Nevertheless, these techniques require more computational burden than the WT-based methods.

4.5. Hybrid Methods

In the present standard [25], it is established that any requirements for ride-through shall not be falsely inhibited by any methods or design features utilized to meet the unintentional islanding detection when an actual unintentional island condition does not exist. Conversely, the unintentional islanding detection requirements shall not be inhibited by ride-through during valid unintentional islanding conditions.

Passive methods operating alone cannot provide satisfying results considering their poor islanding detection performance and the need to ensure ride-through capability by the PVS. Better results can be obtained using active methods, but the power quality is often affected.

Taking into account also the recent advancement in computing capability and communication systems, hybrid islanding detection techniques, based on combinations of the four categories previously discussed, represent the most promising techniques [115–125]. Starting from combination or modification of conventional methods, in [115] some islanding detection methods are proposed as integration of well-known passive methods. In [116] a modified active SVS method is presented, and the modulation index is used as injected signal to achieve the voltage magnitude shift and to detect the islanding phenomenon. In [117] a hybrid active method is developed combining a threshold filter (based on a binary tree classifier) with a harmonic amplification factor used as perturbation to detect islanding. In [118] a modified active islanding detection method is obtained varying the amplitude of the PVS current periodically. In this case islanding occurrence is detected through the AFD method when high current variations are registered. In [119] an active technique is obtained combining the AFD with the SMS. In [120] a new active islanding detection method is proposed and based on the droop control theory. The droop control is modified considering a correlation function between the frequency and the reactive power. This function is used to detect the islanding condition. Moving towards more innovative solutions, in [121] a new islanding detection technique is based on the addition of a variable impedance and a hybrid automatic transfer switch. In [122] computational geometry has been applied to derive an islanding detection technique based on a classifier module. As in case of the active methods, the technique discussed in [123] uses harmonic current injection. The islanding condition is detected through cross correlation. In [124] an original islanding detection scheme based on machine learning adoption is proposed. In [125] a communication-based islanding detection method is presented based on a wireless sensors network. The performances are improved adding a combination of selected loads in the system in order to avoid excessive voltage variations.

4.6. Performances Evaluation of the Islanding Detection Methods

The performances of the considered islanding detection categories are reviewed in Table 1. The main advantages and disadvantages related to the same categories are summarized in Table 2. The hybrid methods are combinations of the islanding detection techniques categorized as communication-based methods, passive methods, active methods and signal processing-based methods. The performances of the hybrid methods depend on the original characteristics of the techniques that they match, hence hybrid methods are not reported in Tables 1 and 2. The hybrid methods are designed to overcome the disadvantages of the islanding detection techniques previously developed. As a consequence, hybrid methods are progressively updated, and they can be assessed as tradeoff among advantages of the original methods and increased implementation complexity.

Table 1. Performances of the islanding detection methods.

	Communication-Based Methods	Passive Methods	Active Methods	Signal Processing-Based Methods
Accuracy	high	poor	high	high
NDZ area	absent	wide	limited	absent or limited
Dynamics	fast	generally fast	medium slow	fast
Impact on the grid	no	no	medium or significant	no
Price	high	limited	limited	high

Table 2. Advantages and disadvantages of the islanding detection methods.

	Communication-Based Methods	Passive Methods	Active Methods	Signal Processing-Based Methods
Advantage	reliability	simplicity	accuracy relative simplicity	accuracy
Disadvantage	expensive with regard to low power PVSs	erroneous detection in case of power balance between generation and load improper estimation due to ride-through requirements compliance	power quality detriment stability hazard in case of numerous PVSs connected in parallel	complexity

5. Synchronization and Islanding Detection Coordination

ZCD methods and ZCD-based PLLs exhibit low dynamic performance and are not suitable for grid voltage monitoring in case of abrupt changes of the grid voltage and power quality disturbances. Arctangent-based PLLs are not particularly widespread due to implementation issues. The PLLs based on Park transform and SOGI OSG, known as SOGI PLLs and the EPLLs represent the most promising synchronization systems for single-phase PVSs due to high filtering capability, also in presence of grid voltage harmonic distortion, accuracy and high dynamic performances also in case of grid voltage abrupt variations. SOGI PLLs and EPLLs are ideal candidates to be employed in the islanding detection and in the reconnection of a PVS to the main grid after an islanding event.

5.1. Impact of the Synchronization Systems on the Islanding Detection Methods

Many islanding detection methods are based on information about the amplitude, the phase and the frequency of the PVS voltage. These methods do not require additional synchronization systems to be included in the PVS control structure. The SOGI PLLs and the EPLLs represent the best synchronization systems for this kind of applications. Other islanding detection techniques require additional synchronization systems in order to monitor some selected harmonics and the PVS current or can require some modification of the synchronization system generally used to track the PVS voltage. There are also some islanding detection techniques which are not based on synchronization systems information. In Table 3 there are reported the main devices used by the different islanding detection methods discussed in Section 4. The islanding detection methods which employ PLLs or EPLLs are pointed out. Hybrid methods are not included in Table 3 since their characteristics depend on the original methods that they combine.

Looking at Table 3, it can be observed that both communication-based techniques and signal processing-based techniques avoid the use of PLLs. Indeed, the communication-based islanding detection techniques are based on communication interface equipment, in particular receivers. The signal processing-based techniques are based on harmonics measurement and localization related to time-varying electrical signals such as voltage, power, entropy, etc. The harmonics decomposition is achieved through the use of the FT, the WT or the ST.

Table 3. Main devices of the islanding detection techniques.

Method		Devices
Communication-based	PLCC-based	Receivers
	SCADA-based	
Passive	OUV/OUF	Solid-state relays or PLLs/EPLLs coordinated with relays
	ROCOF	
	PJ	Additional PLL/EPLL for current monitoring
	VHM	Additional PLLS/EPLLS for voltage harmonics monitoring
Active	Grid impedance variation	Impedance measurement
	Active and reactive power injections	Solid-state relays or PLLs/EPLLs coordinated with relays
	AFD	
	SFS	
	SVS	
	SMS	Modified or additional PLL/EPLL
Signal-processing-based	FT-based	Harmonics measurement and localization of electrical time-varying signals
	WT-based	
	ST-based	

The operating principle of most active and passive methods is based on PLLs or EPLLs. Among the passive methods, the OUV, OUF and ROCOF methods are based on PVS voltage amplitude and frequency information. The PVS PLL or EPLL, used to monitor the PVS operation, can be employed to detect the islanding occurrence operating in coordination with the OUV, OUF and ROCOF relays. Some OUV, OUF and ROCOF protections avoid the PLL and are just based on solid-state relays. The choice depends on the PVS power size, the cost, the desired level of performance. Similarly, many active methods act in order to determine an OUV or OUF in the PVS. When the voltage or the frequency exceeds the threshold value, islanding is detected. It occurs for the active and reactive power injections methods, the AFD, the SFS and the SVS methods. Furthermore, for these methods the main devices are solid-state relays or PLLs/EPLLs coordinated with relays.

Among the passive methods the PJ method requires an additional PLL/EPLL to monitor the PVS current phase which has to be compared to the PVS voltage phase. The VHM requires more PLLS/EPLLs to track the selected harmonics variations and to detect the islanding condition. Among the active methods the SMS method is based on an additional PLL or on a modification of the PVS PLL to create the perturbation in the PVS and to detect the islanding occurrence. In the field of the active methods, just the grid impedance variation method does not require information provided by the PLL/EPLL since it is based on the impedance measurement.

5.2. Reconnection of a PVS to the Grid after an Islanding Event

Independently of the adopted islanding detection technique, the reconnection of a PVS after an islanding event needs to be managed by two PLLS or EPLLs circuits. Indeed, an improper reconnection event is not improbable if the PVS breaker S_1 connects the system to the grid when the PVS voltage is out of phase compared to the grid voltage. In this occasion overcurrents can be verified or, in the worst case, a second disconnection can occur determined by the PVS protections intervention. Hence the reclosing procedure has to be managed in strict coordination with the PVS synchronization system.

Assuming, for example, to detect the islanding condition using the active and reactive power injections method and to employ SOGI PLLs as synchronization systems, in Figure 12 there are shown the PVS voltage e_{PVS} and the grid voltage e in case islanding occurs at $t = 10$ s. At this time, the grid utility breaker Sg is opened and during the islanding operation the amplitude and the frequency of the PVS voltage drift from the rated values.

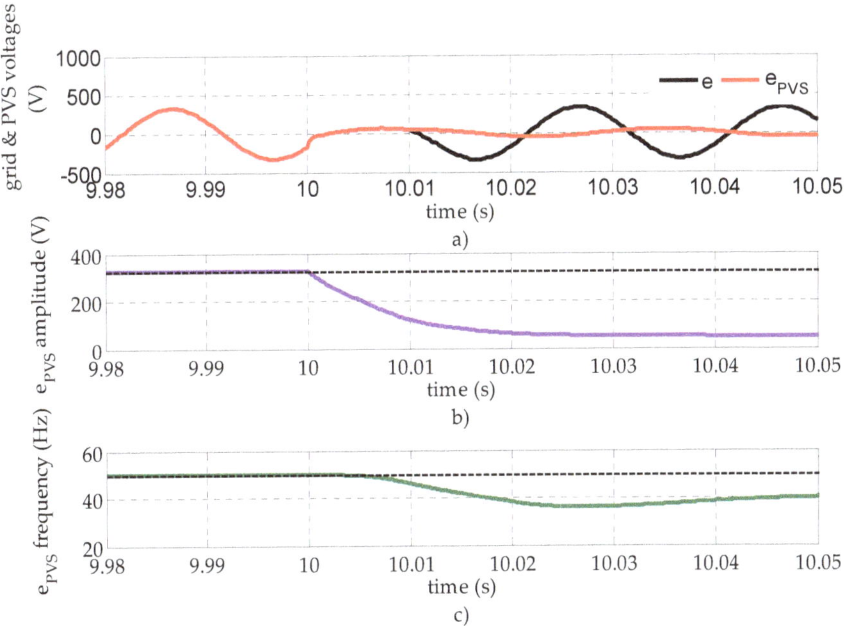

Figure 12. Grid and PVS voltages in case of islanding occurrence. (**a**) e and e_{PVS} waveforms; (**b**) e_{PVS} amplitude; (**c**) e_{PVS} frequency.

The information about the amplitude and the frequency of the PVS voltage, provided by the SOGI PLL, is employed to assess the islanding phenomenon within 2 s on the basis of the active and reactive power injections method. The grid and the PVS voltage waveforms are not more synchronized and, when the e_{PVS} amplitude and frequency deviations exceed the thresholds values, islanding is detected and also the PVS breaker S_1 is opened.

Denoting as *ID* the control signal providing information about the islanding condition, it is possible to define $ID = 1$ when islanding is not detected and $ID = 0$ when islanding is detected. Similarly, it is possible to use a control signal to assess synchronization of the PVS with the grid. In this analysis it is indicated with synchronization = 1 the condition when e_{PVS} and e are in-phase, synchronization = 0 the condition when the two systems are not synchronized.

Assuming that the grid is recovered in few seconds, at $t_1 = 15$ s the grid breaker Sg is closed again (Figure 13). Nevertheless, the reconnection of the PVS cannot be immediate. Since the grid and the PVS are not more synchronized after the islanding occurrence, some reconnection time is required. When the breaker Sg is reclosed, the control signal moves from $ID = 0$ to $ID = 1$. In the considered case study, the grid recovery is detected in less than 0.03 s, but the PVS is reconnected just at $t_2 = 15.25$.

The PVS reconnection is possible only when e_{PVS} and e are assessed again in-phase. In particular, the synchronization is detected when the phase and the amplitude difference between e_{PVS} and e is null. Only at this time the synchronization control signal moves from synchronization = 0 to synchronization = 1.

In the described procedure two PLLs/EPLLs are required: one to monitor the PVS voltage and one to monitor the grid voltage. This example is provided to point out the role of the synchronization system and to demonstrate the need of a strict coordination between the islanding detection and the synchronization control units.

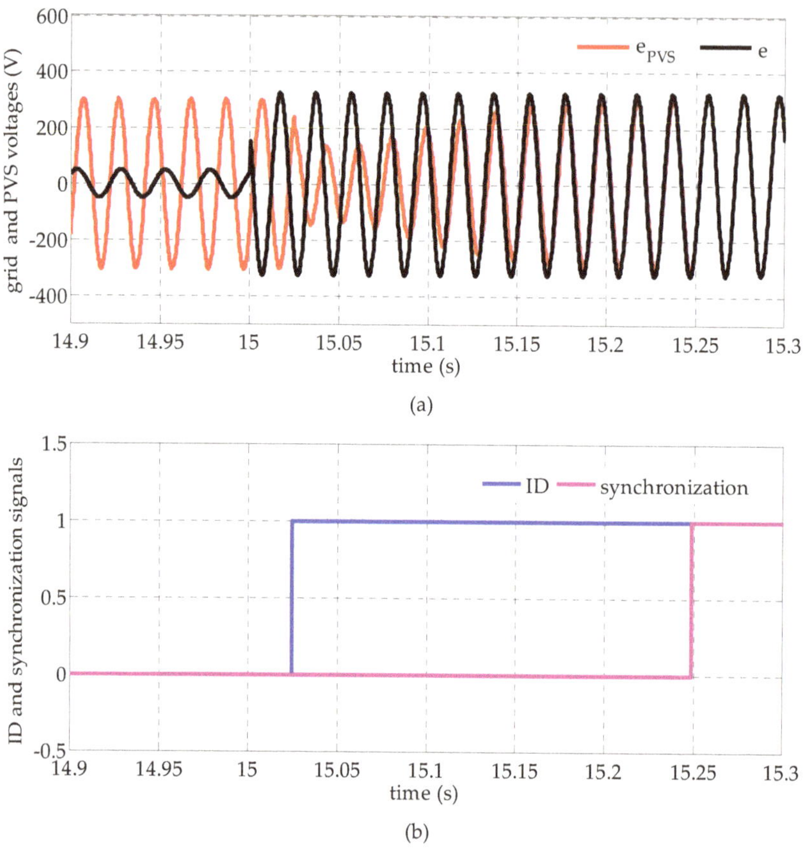

Figure 13. Grid and PVS voltages during resynchronization process. (**a**) e and e_{PVS} waveforms; (**b**) ID and synchronization control signals.

6. Conclusions

An extensive analysis of synchronization and islanding detection methods for single-stage PVSs is presented in this study. Synchronization and islanding detection represent some of the most important control issues for PVSs in the light of the new standards requirements. Abnormal conditions can arise on the utility grid which require a prompt response from the grid-connected PVSs, hence the information provided by the synchronization system are fundamental for the grid voltage monitoring.

Synchronization and islanding detection techniques must operate in coordination. The islanding detection techniques are based also on the information provided by the synchronization techniques. Besides some islanding detection methods use additional PLLs for the harmonics monitoring. In other cases, the normal operation of the PLL is modified in order to detect the islanding phenomenon as it happens in case of the slip-mode frequency shift technique. Finally, it has to be considered that, after a disconnection due to the islanding detection, an improper reconnection event is not improbable if the PVS breaker connects the system to the grid when the PVS voltage is out of phase. In this hypothesis a second disconnection can occur due to the PVS protections action. Hence the reclosing procedure has to be managed with two synchronization systems.

For all these reasons synchronization and islanding detection issues were analyzed together. More than 120 publications were revised and discussed in order to provide a combined review. Both the synchronization and the islanding detection techniques were categorized. The choice of the islanding detection technique depends on numerous criteria. The EPLL and the SOGI PLL represent the preferable synchronization systems to operate in coordination with the islanding detection techniques.

Funding: This research received no external funding.

Conflicts of Interest: The authors declare no conflict of interest.

References

1. IEA PVPS Annual Report 2019. Available online: https://iea-pvps.org/wp-content/uploads/2020/02/5319-iea-pvps-report-2019-08-lr.pdf (accessed on 26 March 2020).
2. Solar PV—Statistics & Facts, The Statistics Portal. Available online: https://www.statista.com/topics/993/solar-pv/ (accessed on 15 April 2020).
3. International Renewable Energy Agency IRENA. Renewable Capacity Statistics 2020. Available online: https://www.irena.org/publications/2020/Mar/Renewable-Capacity-Statistics-2020 (accessed on 26 March 2020).
4. Kausika, B.B.; Moraitis, P.; van Sark, W.G.J.H.M. Visualization of Operational Performance of Grid-Connected PV Systems in Selected European Countries. *Energies* **2018**, *11*, 1330. [CrossRef]
5. Ghosh, S.; Rahman, S. Global deployment of solar photovoltaics: Its opportunities and challenges. In Proceedings of the 2016 IEEE PES Innovative Smart Grid Technologies Conference Europe (ISGT-Europe), Ljubljana, Slovenia, 9–12 October 2016; pp. 1–6. [CrossRef]
6. Araujo, S.V.; Zacharias, P.; Mallwitz, R. Highly Efficient Single-Phase Transformerless Inverters for Grid-Connected Photovoltaic Systems. *IEEE Trans. Ind. Electron.* **2010**, *57*, 3118–3128. [CrossRef]
7. Olowu, T.O.; Sundararajan, A.; Moghaddami, M.; Sarwat, A.I. Future Challenges and Mitigation Methods for High Photovoltaic Penetration: A Survey. *Energies* **2018**, *11*, 1782. [CrossRef]
8. Integrating Renewable Electricity on the Grid—A Report by the APS Panel on Public Affairs, Washington, DC, USA. Available online: https://www.smartgrid.gov/document/integrating_renewable_electricity_grid_report_aps_panel_public_affairs (accessed on 20 November 2011).
9. Ropp, M.; Newmiller, J.; Whitaker, C.; Norris, B. Review of potential problems and utility concerns arising from high penetration levels of photovoltaics in distribution systems. In Proceedings of the 33rd IEEE Photovoltaic Specialists Conference PVSC '08, San Diego, CA, USA, 11–16 May 2008; pp. 1–6. [CrossRef]
10. Appen, J.V.; Braun, M.; Stetz, T.; Diwold, K.; Geibel, D. Time in the Sun: The Challenge of High PV Penetration in the German Electric Grid. *IEEE Power Energy Mag.* **2013**, *11*, 55–64. [CrossRef]
11. Molina-García, A.; Mastromauro, R.A.; García-Sánchez, T.; Pugliese, S.; Liserre, M.; Stasi, S. Reactive Power Flow Control for PV Inverters Voltage Support in LV Distribution Networks. *IEEE Trans. Smart Grid* **2017**, *8*, 447–456. [CrossRef]
12. Vasquez, J.C.; Mastromauro, R.A.; Guerrero, J.M.; Liserre, M. Voltage Support Provided by a Droop-Controlled Multifunctional Inverter. *IEEE Trans. Ind. Electron.* **2009**, *56*, 4510–4519. [CrossRef]
13. Mastromauro, R.A.; Liserre, M.; Kerekes, T.; Dell'Aquila, A. A Single-Phase Voltage-Controlled Grid-Connected Photovoltaic System with Power Quality Conditioner Functionality. *IEEE Trans. Ind. Electron.* **2009**, *56*, 4436–4444. [CrossRef]
14. Koutroulis, E.; Blaabjerg, F. Design Optimization of Transformerless Grid-Connected PV Inverters Including Reliability. *IEEE Trans. Power Electron.* **2013**, *28*, 325–335. [CrossRef]
15. Zhang, L.; Sun, K.; Feng, L.; Wu, H.; Xing, Y. A Family of Neutral Point Clamped Full-Bridge Topologies for Transformerless Photovoltaic Grid-Tied Inverters. *IEEE Trans. Power Electron.* **2013**, *28*, 730–739. [CrossRef]
16. Freddy, T.K.S.; Rahim, N.A.; Hew, W.; Che, H.S. Comparison and Analysis of Single-Phase Transformerless Grid-Connected PV Inverters. *IEEE Trans. Power Electron.* **2014**, *29*, 5358–5369. [CrossRef]
17. Yang, Y.; Blaabjerg, F.; Wang, H. Low-Voltage Ride-Through of Single-Phase Transformerless Photovoltaic Inverters. *IEEE Trans. Ind. Appl.* **2014**, *50*, 1942–1952. [CrossRef]
18. Teodorescu, R.; Liserre, M.; Rodriguez, P. *Grid Converters for Photovoltaic and Wind Power Systems*; IEEE/Wiley: Chichester, UK, 2011.

19. Mastromauro, R.A.; Liserre, M.; Dell'Aquila, A. Control Issues in Single-Stage Photovoltaic Systems: MPPT, Current and Voltage Control. *IEEE Trans. Ind. Inform.* **2012**, *8*, 241–254. [CrossRef]
20. Johnson, B.B.; Dhople, S.V.; Hamadeh, A.O.; Krein, P.T. Synchronization of Parallel Single-Phase Inverters with Virtual Oscillator Control. *IEEE Trans. Power Electron.* **2014**, *29*, 6124–6138. [CrossRef]
21. Hadjidemetriou, L.; Kyriakides, E.; Yang, Y.; Blaabjerg, F. A Synchronization Method for Single-Phase Grid-Tied Inverters. *IEEE Trans. Power Electron.* **2016**, *31*, 2139–2149. [CrossRef]
22. Shitole, A.B.; Suryawanshi, H.M.; Talapur, G.G.; Sathyan, S.; Ballal, M.S.; Borghate, V.B.; Ramteke, M.R.; Chaudhari, M.A. Grid Interfaced Distributed Generation System with Modified Current Control Loop Using Adaptive Synchronization Technique. *IEEE Trans. Ind. Inform.* **2017**, *13*, 2634–2644. [CrossRef]
23. Nagliero, A.; Mastromauro, R.A.; Liserre, M.; Dell'Aquila, A. Monitoring and Synchronization Techniques for Single-Phase PV Systems. In Proceedings of the 2010 International Symposium on Power Electronics, Electrical Drives, Automation and Motion SPEEDAM 2010, Pisa, Italy, 14–16 June 2010; pp. 1404–1409. [CrossRef]
24. Ghartemani, M.K.; Iravani, M.R. A nonlinear adaptative filter for online signal analysis in power systems applications. *IEEE Trans. Power Deliv.* **2002**, *17*, 617–622. [CrossRef]
25. *IEEE Standard for Interconnection and Interoperability of Distributed Energy Resources with Associated Electric Power Systems Interfaces*; IEEE Std 1547-2003; IEEE: Piscataway, NJ, USA, 6 April 2018; pp. 1–138. [CrossRef]
26. Anani, N.; AlAli, O.A.-K.; Al-Qutayri, M.; AL-Araji, S. Synchronization of a renewable energy inverter with the grid. *J. Renew. Sustain. Energy* **2012**, *4*. [CrossRef]
27. Lubura, S.; Soja, M.; Lale, S.; Ikić, M. Single-phase phase locked loop with dc offset and noise rejection for photovoltaic inverters. *IET Power Electron.* **2014**, *7*, 2288–2299. [CrossRef]
28. Luna, A.; Rocabert, J.; Candela, J.I.; Hermoso, J.R.; Teodorescu, R.; Blaabjerg, F.; Rodríguez, P. Grid Voltage Synchronization for Distributed Generation Systems Under Grid Fault Conditions. *IEEE Trans. Ind. Appl.* **2015**, *51*, 3414–3425. [CrossRef]
29. Bower, W.; Ropp, M. *Evaluation of Islanding Detection Methods for Utility-Interactive Inverters in Photovoltaic Systems*. SANDIA REPORT SAND2002-3591. November 2002. Available online: https://pdfs.semanticscholar.org/2a29/ad6772ffef963d7dfdaece6a9db374e5e3b6.pdf (accessed on 29 March 2020).
30. Zhou, Y.; Li, H.; Liu, L. Integrated Autonomous Voltage Regulation and Islanding Detection for High Penetration PV Applications. *IEEE Trans. Power Electron.* **2013**, *28*, 2826–2841. [CrossRef]
31. Yang, F.; Xia, N.; Han, Q. Event-Based Networked Islanding Detection for Distributed Solar PV Generation Systems. *IEEE Trans. Ind. Inform.* **2017**, *13*, 322–329. [CrossRef]
32. Baghaee, H.R.; Mlakić, D.; Nikolovski, S.; Dragicčvić, T. Anti-Islanding Protection of PV-Based Microgrids Consisting of PHEVs Using SVMs. *IEEE Trans. Smart Grid* **2020**, *11*, 483–500. [CrossRef]
33. Reddy, V.R.; Sreeraj, S.E. A Feedback-Based Passive Islanding Detection Technique for One-Cycle-Controlled Single-Phase Inverter Used in Photovoltaic Systems. *IEEE Trans. Ind. Electron.* **2020**, *67*, 6541–6549. [CrossRef]
34. Hung, G.-K.; Chang, C.-C.; Chen, C.-L. Automatic phase-shift method for islanding detection of grid-connected photovoltaic inverters. *IEEE Trans. Energy Convers.* **2003**, *18*, 169–173. [CrossRef]
35. Serban, E.; Serban, H. A Control Strategy for a Distributed Power Generation Microgrid Application with Voltage- and Current-Controlled Source Converter. *IEEE Trans. Power Electron.* **2010**, *25*, 2981–2992. [CrossRef]
36. *IEEE Application Guide for IEEE Std 1547™, IEEE Standard for Interconnecting Distributed Resources with Electric Power Systems*; IEEE: Piscataway, NJ, USA, April 2009; pp. 1–217. [CrossRef]
37. *IEEE Approved Draft Standard Conformance Test Procedures for Equipment Interconnecting Distributed Energy Resources with Electric Power Systems and Associated Interfaces*; IEEE P1547.1/D9.9; IEEE: Piscataway, NJ, USA, January 2020; pp. 1–283.
38. *IEEE Recommended Practice for Utility Interface of Photovoltaic (PV) Systems*; IEEE Std 929-2000; IEEE: New York, NY, USA, 2000. [CrossRef]
39. IEC 61727-2004. *Photovoltaic (PV) Systems—Characteristics of the Utility Interface*; International Electrotechnical Commission: Geneva, Switzerland, 2004.
40. GB/T 19939-2005. *Technical Requirements for Grid Connection of PV System*; China National Standardization Administration Committee: Beijing, China, 2005.
41. CIGRE Working Group B5. *The Impact of Renewable Energy Sources and Distributed Generation on Substation Protection and Automation*; CIGRE: Paris, France, 2010.

42. Gardner, F.M. *Phaselock Techniques*, 2nd ed.; Wiley-Interscience: Hoboken, NJ, USA, 1979; p. 304. ISBN -10: 0471042943.

43. Jaalam, N.; Rahim, N.A.; Bakar, A.H.A.; Tan, C.; Haidar, A.M.A. A comprehensive review of synchronization methods for grid-connected converters of renewable energy source. *Renew. Sustain. Energy Rev.* **2016**, *59*, 1471–1481. [CrossRef]

44. Guan-Chyun, H.; Hung, J.C. Phase-locked loop techniques. A Survey. *IEEE Trans. Ind. Electron.* **1996**, *43*, 609–615. [CrossRef]

45. Golestan, S.; Guerrero, J.M.; Vasquez, J.C. Three-Phase PLLs: A Review of Recent Advances. *IEEE Trans. Power Electron.* **2017**, *32*, 1894–1907. [CrossRef]

46. Golestan, S.; Monfared, M.; Freijedo, F. Design-oriented study of advanced synchronous reference frame phase-locked loops. *IEEE Trans. Power Electron.* **2013**, *28*, 765–778. [CrossRef]

47. Paiva, S.C.; Sanca, H.S.; Costa, F.B.; Souza, B.A. Reviewing of anti-islanding protection. In Proceedings of the 11th IEEE/IAS International Conference on Industry Applications, Juiz de Fora, Brazil, 7–10 December 2014; pp. 1–8. [CrossRef]

48. Li, C.; Cao, C.; Cao, Y.; Kuang, Y.; Zeng, L.; Fang, B. A review of islanding detection methods for microgrid. *Renew. Sustain. Energy Rev.* **2014**, *35*, 211–220. [CrossRef]

49. de Mango, F.; Liserre, M.; Dell'Aquila, A.; Pigazo, A. Overview of anti-islanding algorithms for PV systems. Part I: Passive methods. In Proceedings of the 12th International Power Electronics and Motion Conference, Portoroz, Slovenia, 30 August–1 September 2006; pp. 1878–1883. [CrossRef]

50. de Mango, F.; Liserre, M.; Dell'Aquila, A. Overview of anti-islanding algorithms for PV systems. Part II: Active methods. In Proceedings of the 12th International Power Electronics and Motion Conference, Portoroz, Slovenia, 30 August 30–1 September 2006; pp. 1884–1889. [CrossRef]

51. Abokhalil, A.G.; Awan, A.B.; Al-Qawasmi, A.R. Comparative Study of Passive and Active Islanding Detection Methods for PV Grid-Connected Systems. *Sustainability* **2018**, *10*, 1798. [CrossRef]

52. Mahat, P.; Chen, Z.; Bak-Jensen, B. Review of islanding detection methods for distributed generation. In Proceedings of the Third International Conference on Electric Utility Deregulation and Restructuring and Power Technologies, Nanjing, China, 6–8 April 2008; pp. 2743–2748. [CrossRef]

53. Estébanez, E.J.; Moreno, V.M.; Pigazo, A.; Liserre, M.; Dell'Aquila, A. Performance Evaluation of Active Islanding-Detection Algorithms in Distributed-Generation Photovoltaic Systems: Two Inverters Case. *IEEE Trans. Ind. Electron.* **2011**, *58*, 1185–1193. [CrossRef]

54. Trujillo, C.L.; Velasco, D.; Figueres, E.; Garcerá, G. Analysis of active islanding detection methods for grid-connected microinverters for renewable energy processing. *Appl. Energy* **2010**, *87*, 3591–3605. [CrossRef]

55. Raza, S.; Mokhlis, H.; Arof, H.; Laghari, J.A.; Wang, L. Application of signal processing techniques for islanding detection of distributed generation in distribution network: A review. *Energy Convers. Manag.* **2015**, *96*, 613–624. [CrossRef]

56. Manikonda, S.K.G.; Gaonkar, D.N. Comprehensive review of IDMs in DG systems. *IET Smart Grid* **2019**, *2*, 11–24. [CrossRef]

57. Khamis, A.; Shareef, H.; Bizkevelci, E.; Khatib, T. A review of islanding detection techniques for renewable distributed generation systems. *Renew. Sustain. Energy Rev.* **2013**, *28*, 483–493. [CrossRef]

58. Singh, S.; Kewat, S.; Singh, B.; Panigrahi, B.K.; Kushwaha, M.K. Seamless Control of Solar PV Grid Interfaced System with Islanding Operation. *IEEE Power Energy Technol. Syst. J.* **2019**, *6*, 162–171. [CrossRef]

59. Lee, K.J.; Lee, J.P.; Shin, D.; Yoo, D.W.; Kim, H.J. A novel grid synchronization pll method based on adaptive low-pass notch filter for grid-connected pcs. *IEEE Trans. Ind. Electron.* **2014**, *61*, 292–301. [CrossRef]

60. Wang, Y.F.; Li, Y.W. A Grid Fundamental and Harmonic Component Detection Method for Single-Phase Systems. *IEEE Trans. Power Electron.* **2013**, *28*, 2204–2213. [CrossRef]

61. Karimi-Ghartemani, M. A Unifying Approach to Single-Phase Synchronous Reference Frame PLLs. *IEEE Trans. Power Electron.* **2013**, *28*, 4550–4556. [CrossRef]

62. Wall, R.W. Simple methods for detecting zero crossing. In Proceedings of the 29th Annual Conference of the IEEE Industrial Electronics Society IECON'03, Roanoke, VA, USA, 2–6 November 2003; Volume 3, pp. 2477–2481. [CrossRef]

63. Golestan, S.; Monfared, M.; Freijedo, F.D.; Guerrero, J.M. Dynamics Assessment of Advanced Single-Phase PLL Structures. *IEEE Trans. Ind. Electron.* **2013**, *60*, 2167–2177. [CrossRef]

64. Kandeepan, S.; Reisenfeld, S. Frequency tracking and acquisition with a four-quadrant arctan phase detector based digital phase locked loop. In Proceedings of the 2003 Joint Fourth International Conference on Information, Communications and Signal Processing and the Fourth Pacific Rim Conference on Multimedia, Singapore, 15–18 December 2003; Volume 1, pp. 401–405. [CrossRef]

65. Konara, K.M.S.Y.; Kolhe, M.L.; Sankalpa, W.G.C.A. Grid synchronization of DC energy storage using Voltage Source Inverter with ZCD and PLL techniques. In Proceedings of the 2015 IEEE 10th International Conference on Industrial and Information Systems (ICIIS), Peradeniya, Sri Lanka, 18–20 December 2015; pp. 458–462. [CrossRef]

66. Weidenbrug, R.; Dawson, F.P.; Bonert, R. New synchronization method for thyristor power converters to weak AC-systems. *IEEE Trans. Ind. Electron.* **1993**, *40*, 505–511. [CrossRef]

67. Mur, F.; Cardenas, V.; Vaquero, J.; Martinez, S. Phase synchronization and measurement digital systems of AC mains for power converters. In Proceedings of the 6th IEEE Power Electronics Congress CIEP 98, Morelia, Mexico, 12–15 October 1998; pp. 188–194. [CrossRef]

68. Choi, J.-W.; Kim, Y.-K.; Kim, H.-G. Digital PLL control for single-phase photovoltaic system. *IEE Proc. EE Electr. Power Appl.* **2006**, *153*, 40–46. [CrossRef]

69. Gonzalez-Espin, F.; Figueres, E.; Garcera, G. An Adaptive Synchronous-Reference-Frame Phase-Locked Loop for Power Quality Improvement in a Polluted Utility Grid. *IEEE Trans. Ind. Electron* **2012**, *59*, 2718–2731. [CrossRef]

70. Yazdani, D.; Bakhshai, A.; Jain, P. Grid synchronization techniques for converter interfaced distributed generation systems. In Proceedings of the 2009 Energy Conversion Congress and Exposition, ECCE 2009, San Jose, CA, USA, 20–24 September 2009; pp. 2007–2014. [CrossRef]

71. Han, Y.; Luo, M.; Zhao, X.; Guerrero, J.M.; Xu, L. Comparative Performance Evaluation of Orthogonal-Signal-Generators-Based Single-Phase PLL Algorithms—A Survey. *IEEE Trans. Power Electron.* **2016**, *31*, 3932–3944. [CrossRef]

72. Yang, Y.; Blaabjerg, F.; Zou, Z. Benchmarking of Grid Fault Modes in Single-Phase Grid-Connected Photovoltaic Systems. *IEEE Trans. Ind. Appl.* **2013**, *49*, 2167–2176. [CrossRef]

73. Rodriguez, P.; Teodorescu, R.; Candela, I.; Timbus, A.V.; Laabjerg, F.B. New positive-sequence voltage detector for grid synchronization of power converters under faulty grid conditions. In Proceedings of the 2006 37th IEEE Power Electronics Specialists Conference PESC'06, Jeju, Korea, 18–22 June 2006; pp. 1–7. [CrossRef]

74. Chittora, P.; Singh, A.; Singh, M. Adaptive EPLL for improving power quality in three-phase three-wire grid-connected photovoltaic system. *IET Renew. Power Gener.* **2019**, *13*, 1595–1602. [CrossRef]

75. Karimi-Ghartemani, M. *Enhanced Phase-Locked Loop Structures for Power and Energy Applications*; Wiley-IEEE Press: Piscataway, NJ, USA, 21 March 2014. [CrossRef]

76. Karimi-Ghartemani, M. Linear and Pseudolinear Enhanced Phased-Locked Loop (EPLL) Structures. *IEEE Trans. Ind. Electron.* **2014**, *61*, 1464–1474. [CrossRef]

77. Sahoo, A.; Mahmud, K.; Ciobotaru, M.; Ravishankar, J. Adaptive Grid Synchronization Technique for Single-phase Inverters in AC Microgrid. In Proceedings of the 2019 IEEE Energy Conversion Congress and Exposition (ECCE), Baltimore, MD, USA, 29 September–3 October 2019; pp. 4441–4446. [CrossRef]

78. Task V Report IEA-PVPS T5-09:2002, Evaluation of Islanding Detection Methods for Photovoltaic Utility-Interactive Power Systems. Available online: https://iea-pvps.org/wp-content/uploads/2020/01/rep5_09.pdf (accessed on 15 December 2002).

79. Ye, Z.; Kolwalkar, A.; Zhang, Y.; Du, P.; Walling, R. Evaluation of anti-islanding schemes based on nondetection zone concept. *IEEE Trans. Power Electron.* **2004**, *19*, 1171–1176. [CrossRef]

80. Ropp, M.E.; Begovic, M.; Rohatgi, A.; Kern, G.A.; Bonn, R.H.; Gonzalez, S. Determining the relative effectiveness of islanding detection methods using phase criteria and nondetection zones. *IEEE Trans. Energy Convers.* **2000**, *15*, 290–296. [CrossRef]

81. *IEEE Standard for Interconnecting Distributed Resources with Electric Power Systems*; IEEE Std 1547-2003; IEEE: Piscataway, NJ, USA, 1–28 July 2003. [CrossRef]

82. Shrestha, A.; Kattel, R.; Dachhepatic, M.; Mali, B.; Thapa, R.; Singh, A.; Bista, D.; Adhikary, B.; Papadakis, A.; Maskey, R.K. Comparative Study of Different Approaches for Islanding Detection of Distributed Generation Systems. *Appl. Syst. Innov.* **2019**, *2*. [CrossRef]

83. Samuelsson, O.; Strath, N. Islanding detection and connection requirements. In Proceedings of the 2007 IEEE Power Engineering Society General Meeting, Tampa, FL, USA, 24–28 June 2007; pp. 1–6. [CrossRef]

84. Ropp, M.; Larson, D.; Meendering, S.; MacMahon, D.; Ginn, J.; Stevens, J.; Bower, W.; Gonzalez, S.; Fennell, K.; Brusseau, L. Discussion of power line carrier communications-based anti-islanding scheme using a commercial automatic meter reading system. In Proceedings of the 4th IEEE World Conference on Photovoltaic Energy Conversion, Waikoloa, HI, USA, 7–12 May 2006; pp. 2351–2354. [CrossRef]

85. Benato, R.; Caldon, R. Distribution line carrier: Analysis procedure and applications to DG. *IEEE Trans. Power Deliv.* **2007**, *22*, 575–583. [CrossRef]

86. Xu, W.; Zhang, G.; Li, C.; Wang, W.; Wang, G.; Kliber, J. A power line signaling based technique for anti-islanding protection of distributed generators—Part i: Scheme and analysis. *IEEE Trans. Power Deliv.* **2007**, *22*, 1758–1766. [CrossRef]

87. Wang, G.; Kliber, J.; Zhang, G.; Xu, W.; Howell, B.; Palladino, T. A power line signaling based technique for anti-islanding protection of distributed generators—Part ii: Field test results. *IEEE Trans. Power Deliv.* **2007**, *22*, 1767–1772. [CrossRef]

88. Aphrodis, N.; Ntagwirumugara, E.; Vianney, B.J.M.; Mulolani, F. Design, Control and Validation of a PV System Based on Supervisory Control and Data Acquisition (SCADA) Viewer in Smartgrids. In Proceedings of the 5th International Conference on Control, Automation and Robotics (ICCAR), Beijing, China, 19–22 April 2019; pp. 23–28. [CrossRef]

89. Etxegarai, A.; Egua, P.; Zamora, I. Analysis of remote islanding detection methods for distributed resources. In Proceedings of the International Conference on Renewable Energies and Power Quality (ICREPQ'11), Las Palmas, Spain, 13–15 April 2011; pp. 1142–1147. Available online: http://www.icrepq.com/icrepq\T1\textquoteright11/580-etxegarai.pdf (accessed on 29 March 2020).

90. Vieira, W.X.J.C.; Freitas, W.; Morelato, A. Performance of frequency relays for distributed generation protection. *IEEE Trans. Power Deliv.* **2006**, *21*, 1120–1127. [CrossRef]

91. Freitas, W.; Xu, W.; Affonso, C.M.; Huang, Z. Comparative analysis between ROCOF and vector surge relays for distributed generation applications. *IEEE Trans. Power Deliv.* **2005**, *20*, 1315–1324. [CrossRef]

92. Ten, C.F.; Crossley, P.A. Evaluation of Rocof Relay Performances on Networks with Distributed Generation. In Proceedings of the 2008 IET 9th International Conference on Developments in Power System Protection (DPSP 2008), Glasgow, Scotland, 17–20 March 2008; pp. 523–528. [CrossRef]

93. de la Fuente, D.V.; Rodriguez, C.L.T.; Narvaez, E.A. Review of Anti-Islanding Methods: Analysis by Figures of Merit Tools for Controllers Reconfiguration in Microgrids. *IEEE Lat. Am. Trans.* **2015**, *13*, 679–686. [CrossRef]

94. Timbus, A.V.; Teodorescu, R.; Blaabjerg, F.; Borup, U. Online grid measurement and ENS detection for PV inverter running on highly inductive grid. *IEEE Power Electron. Lett.* **2004**, *2*, 77–82. [CrossRef]

95. Asiminoaei, L.; Teodorescu, R.; Blaabjerg, F.; Borup, U. A digital controlled PV-inverter with grid impedance estimation for ENS detection. *IEEE Trans. Power Electron.* **2005**, *20*, 1480–1490. [CrossRef]

96. Jeraputra, P.N. Enjeti, Development of a robust anti-islanding algorithm for utility interconnection of distributed fuel cell powered generation. *IEEE Trans. Power Electron.* **2004**, *19*, 1163–1170. [CrossRef]

97. Du, P.; Ye, Z.; Aponte, E.E.; Nelson, J.K.; Fan, L. Positive-Feedback-Based Active Anti-Islanding Schemes for Inverter-Based Distributed Generators: Basic Principle, Design Guideline and Performance Analysis. *IEEE Trans. Power Electron.* **2010**, *25*, 2941–2948. [CrossRef]

98. Zeineldin, H.H.; Conti, S. Sandia frequency shift parameter selection for multi-inverter systems to eliminate non-detection zone. *IET Renew. Power Gener.* **2011**, *5*, 175–183. [CrossRef]

99. Reis, M.V.G.; Barros, T.A.S.; Moreira, A.B.; Nascimento, S.F.P.; Ruppert, F.E.; Villalva, M.G. Analysis of the Sandia Frequency Shift (SFS) islanding detection method with a single-phase photovoltaic distributed generation system. In Proceedings of the 2015 IEEE PES Innovative Smart Grid Technologies Latin America ISGT LATAM 2015, Montevideo, Uruguay, 5–7 October 2015; pp. 125–129. [CrossRef]

100. El-Moubarak, M.; Hassan, M.; Faza, A. Performance of three islanding detection methods for grid-tied multi-inverters. In Proceedings of the 2015 IEEE 15th International Conference on Environment and Electrical Engineering (EEEIC), Rome, Italy, 10–13 June 2015; pp. 1999–2004. [CrossRef]

101. Mohammadpour, B.; Zareie, M.; Eren, S.; Pahlevani, M. Stability analysis of the slip mode frequency shift islanding detection in single phase PV inverters. In Proceedings of the 2017 IEEE 26th International Symposium on Industrial Electronics (ISIE), Edinburgh, UK, 19–21 June 2017; pp. 873–878. [CrossRef]

102. Mohammadpour, B.; Pahlevani, M.; Kaviri, S.M.; Jain, P. Advanced slip mode frequency shift islanding detection method for single phase grid connected PV inverters. In Proceedings of the 2016 IEEE Applied Power Electronics Conference and Exposition (APEC), Long Beach, CA, USA, 20–24 March 2016; pp. 378–385. [CrossRef]

103. Gu, Y.H.; Bollen, M.H.J. Time-Frequency and Time-Scale Domain Analysis of Voltage Disturbances. *IEEE Trans. Power Deliv.* **2000**, *15*, 1279–1284. [CrossRef]

104. Kim, I.-S. Islanding Detection Technique using Grid-Harmonic Parameters in the Photovoltaic System. *Energy Procedia* **2012**, *14*, 137–141. [CrossRef]

105. Kim, J.; Kim, J.; Ji, Y.; Jung, Y.; Won, C. An Islanding Detection Method for a Grid-Connected System Based on the Goertzel Algorithm. *IEEE Trans. Power Electron.* **2011**, *26*, 1049–1055. [CrossRef]

106. Hanif, M.; Basu, M.; Gaughan, K. Development of EN50438 compliant wavelet-based islanding detection technique for three-phase static distributed generation systems. *IET Renew. Power Gener.* **2012**, *6*, 289–301. [CrossRef]

107. Zhu, Y.; Yang, Q.; Wu, J.; Zheng, D.; Tian, Y. A novel islanding detection method of distributed generator based on wavelet transform. In Proceedings of the 2008 International Conference on Electrical Machines and Systems, Wuhan, China, 17–20 October 2008; pp. 2686–2688.

108. Ning, J.; Wang, C. Feature extraction for islanding detection using Wavelet Transform-based Multi-Resolution Analysis. In Proceedings of the 2012 IEEE Power and Energy Society General Meeting, San Diego, CA, USA, 22–26 July 2012; pp. 1–6. [CrossRef]

109. Hsieh, C.-T.; Lin, J.-M.; Huang, S.-J. Enhancement of islanding-detection of distributed generation systems via wavelet transform-based approaches. *Int. J. Electr. Power Energy Syst.* **2008**, *30*, 575–580. [CrossRef]

110. Do, H.T.; Zhang, X.; Nguyen, N.V.; Li, S.S.; Chu, T.T. Passive-Islanding Detection Method Using the Wavelet Packet Transform in Grid-Connected Photovoltaic Systems. *IEEE Trans. Power Electron.* **2016**, *31*, 6955–6967. [CrossRef]

111. Pigazo, A.; Liserre, M.; Mastromauro, R.A.; Moreno, V.M.; Dell'Aquila, A. Wavelet-Based Islanding Detection in Grid-Connected PV Systems. *IEEE Trans. Ind. Electron.* **2009**, *56*, 4445–4455. [CrossRef]

112. Ray, P.; Mohanty, S.; Kishor, S.; Dubey, H. Coherency determination in grid connected distributed generation based hybrid system under islanding scenarios. In Proceedings of the IEEE International Conference on Power and Energy (PECon2010), Kuala Lumpur, Malaysia, 29 November–1 December 2010; pp. 85–88. [CrossRef]

113. Ray, P.; Kishor, N.; Mohanty, S. S-Transform based islanding detection in grid connected distributed generation based power system. In Proceedings of the 2010 IEEE International Energy Conference (ENERGYCON2010), Manama, Bahrain, 18–22 December 2010; pp. 612–617. [CrossRef]

114. Ray, P.; Mohanty, S.; Kishor, N. Disturbance detection in grid-connected distributed generation system using wavelet and S-Transform. *Electr. Power Syst. Res.* **2011**, *81*, 805–819. [CrossRef]

115. Quoc-Tuan, T. New methods of islanding detection for photovoltaic inverters. In Proceedings of the 2016 IEEE PES Innovative Smart Grid Technologies Conference Europe (ISGT-Europe), Ljubljana, Slovenia, 9–12 October 2016; pp. 1–5. [CrossRef]

116. Liu, S.; Zhuang, S.; Xu, Q.; Xiao, J. Improved voltage shift islanding detection method for multi-inverter grid-connected photovoltaic systems. *IET Gener. Transm. Distrib.* **2016**, *10*, 3163–3169. [CrossRef]

117. Dhar, S.; Dash, P.K. Harmonic Profile Injection-Based Hybrid Active Islanding Detection Technique for PV-VSC-Based Microgrid System. *IEEE Trans. Sustain. Energy* **2016**, *7*, 1473–1481. [CrossRef]

118. Yu, B.; Matsui, M.; So, J.; Yu, G. A high power quality anti-islanding method using effective power variation. *Sol. Energy* **2008**, *82*, 368–378. [CrossRef]

119. Yu, B.; Matsui, M.; Jung, Y.; Yu, G. A combined active anti-islanding method for photovoltaic systems. *Renew. Energy* **2008**, *33*, 979–985. [CrossRef]

120. Muñoz-Cruzado-Alba, J.; Villegas-Núñez, J.; Vite-Frías, J.A.; Carrasco-Solís, J.M.; Galván-Díez, E. New Low-Distortion Q–f Droop Plus Correlation Anti-Islanding Detection Method for Power Converters in Distributed Generation Systems. *IEEE Trans. Ind. Electron.* **2015**, *62*, 5072–5081. [CrossRef]

121. Papadimitriou, C.N.; Kleftakis, V.A.; Hatziargyriou, N.D. A novel islanding detection method for microgrids based on variable impedance insertion. *Electr. Power Syst. Res.* **2015**, *121*, 58–66. [CrossRef]

122. Vyas, S.; Kumar, R. Computational geometry-based methodology for identification of potential islanding initiators in high solar PV penetration distribution feeders. *IET Renew. Power Gener.* **2018**, *12*, 456–462. [CrossRef]
123. Voglitsis, D.; Papanikolaou, N.P.; Kyritsis, A.C. Active Cross-Correlation Anti-Islanding Scheme for PV Module-Integrated Converters in the Prospect of High Penetration Levels and Weak Grid Conditions. *IEEE Trans. Power Electron.* **2019**, *34*, 2258–2274. [CrossRef]
124. Baghaee, H.R.; Mlakić, D.; Nikolovski, S.; Dragičević, T. Support Vector Machine-based Islanding and Grid Fault Detection in Active Distribution Networks. *IEEE J. Emerg. Sel. Top. Power Electron.* **2019**. [CrossRef]
125. Eshraghi, A.; Ghorbani, R. Islanding detection and over voltage mitigation using controllable loads. *Sustain. Energy Grids Netw.* **2016**, *6*, 125–135. [CrossRef]

MDPI

St. Alban-Anlage 66

4052 Basel

Switzerland

Tel. +41 61 683 77 34

Fax +41 61 302 89 18

www.mdpi.com

Energies Editorial Office

E-mail: energies@mdpi.com

www.mdpi.com/journal/energies